Lecture Notes in Control and Information Sciences

Edited by A. V. Balakrishnan and M. Thoma

Vol. 22: Optimization Techniques
Proceedings of the 9th IFIP Conference on Optimization Techniques,
Warsaw, September 4–8, 1979
Part 1
Edited by K. Iracki, K. Malanowski, S. Walukiewicz
XVI, 569 pages. 1980

Vol. 23: Optimization Techniques
Proceedings of the 9th IFIP Conference on Optimization Techniques,
Warsaw, September 4-8, 1979
Part 2
Edited by K. Iracki, K. Malanowski, S. Walukiewicz
XV, 621 pages. 1980

Vol. 24: Methods and Applications in Adaptive Control
Proceedings of an International Symposium
Bochum, 1980
Edited by H. Unbehauen
VI, 309 pages. 1980

Vol. 25: Stochastic Differential Systems – Filtering and Control
Proceedings of the IFIP-WG7/1 Working Conference
Vilnius, Lithuania, USSR, Aug. 28 – Sept. 2, 1978
Edited by B. Grigelionis
X, 362 pages. 1980

Vol. 26: D. L. Iglehart, G. S. Shedler
Regenerative Simulation of Response Times in Networks of Queues
XII, 204 pages. 1980

Vol. 27: D. H. Jacobson, D. H. Martin, M. Pachter, T. Geveci
Extensions of Linear-Quadratic Control Theory
XI, 288 pages. 1980

Vol. 28: Analysis and Optimization of Systems
Proceedings of the Fourth International Conference on Analysis and Optimization of Systems
Versailles, December 16–19, 1980
Edited by A. Bensoussan and J. L. Lions
XIV, 999 pages. 1980

Vol. 29: M. Vidyasagar,
Input-Output Analysis of Large-Scale Interconnected Systems – Decomposition, Well-Posedness and Stability
VI, 221 pages. 1981

Vol. 30: Optimization and Optimal Control
Proceedings of a Conference Held at Oberwolfach, March 16–22, 1980
Edited by A. Auslender, W. Oettli, and J. Stoer
VIII, 254 pages. 1981

Vol. 31: Berc Rustem
Projection Methods in Constrained Optimisation and Applications to Optimal Policy Decisions
XV, 315 pages. 1981

Vol. 32: Tsuyoshi Matsuo,
Realization Theory of Continuous-Time Dynamical Systems
VI, 329 pages, 1981

Vol. 33: Peter Dransfield
Hydraulic Control Systems – Design and Analysis of Their Dynamics
VII, 227 pages, 1981

Vol. 34: H.W. Knobloch
Higher Order Necessary Conditions in Optimal Control Theory
V, 173 pages, 1981

Vol. 35: Global Modelling
Proceedings of the IFIP-WG 7/1 Working Conference Dubrovnik, Yugoslavia,
Sept. 1–5, 1980
Edited by S. Krčevinac
VIII, 232 pages, 1981

Vol. 36: Stochastic Differential Systems
Proceedings of the 3rd IFIP-WG 7/1 Working Conference
Visegrád, Hungary, Sept. 15–20, 1980
Edited by M. Arató, D. Vermes, A.V. Balakrishn
VI, 238 pages, 1981

Vol. 37: Rüdiger Schmidt
Advances in Nonlinear Parameter Optimization
VI, 159 pages, 1982

Vol. 38: System Modeling and Optimization
Proceedings of the 10th IFIP Conference
New York City, USA, Aug. 31 – Sept. 4, 1981
Edited by R.F. Drenick and F. Kozin
XI, 894 pages. 1982

Vol. 39: Feedback Control of Linear and Nonlinear Systems
Proceedings of the Joint Workshop on Feedback and Synthesis of Linear and Nonlinear Systems
Bielefeld/Rom
XIII, 284 pages. 1982

Vol. 40: Y.S. Hung, A.G.J. MacFarlane
Multivariable Feedback: A Quasi-Classical Approach
X, 182 pages. 1982

Vol. 41: M. Gössel
Nonlinear Time-Discrete Systems – A General Approach by Nonlinear Superposition
VIII, 112 pages. 1982

Vol. 42: Advances in Filtering and Optimal Stochastic Control
Proceedings of the IFIP-WG 7/1 Working Conference
Cocoyoc, Mexico, February 1–6, 1982
VIII, 391 pages. 1982

For information about Vols. 1–21 please contact your bookseller or Springer-Verlag.

Lecture Notes in Control and Information Sciences

Edited by A.V. Balakrishnan and M. Thoma

64

Arunabha Bagchi

Stackelberg Differential Games in Economic Models

Springer-Verlag Berlin Heidelberg GmbH 1984

Author
Arunabha Bagchi
Dept. of Applied Mathematics
Twente University of Technology
P. O. Box 217
7500 AE Enschede
The Netherlands

ISBN 978-3-540-13587-6 ISBN 978-3-540-39021-3 (eBook)
DOI 10.1007/978-3-540-39021-3

Library of Congress Cataloging in Publication Data
Stackelberg differential games in economic models.
(Lecture notes in control and information sciences ; 64)
1. Economics--Mathematical models.
2. Game theory.
3. Differential games.
I. Bagchi, Arunabha.
II. Series.
HB144.S69 1984 330'.0724 84-10606

Originally published by Springer-Verlag Berlin Heidelberg New York in 1984.
Offsetprinting: Mercedes-Druck, Berlin

2061/3020-543210

To

Arundhati and Mishtu

with love

PREFACE

Towards the end of nineteen seventies, I found myself in the scientific company of two experts on Differential Games, Tamer Başar and Geert Jan Olsder. It became impossible to ignore the subject any longer. The result was fruitful collaboration with both of them on several problems in stochastic differential games. My general interest in political economy, however, soon tempted me to formulate some economic models in the framework of Stackelberg (leader-follower) games. This research monograph is an outgrowth of those adventures.

Some applications of Stackelberg differential games to economic models have appeared during the last few years. The purpose of this book is not to write a survey on them but, rather, to concentrate mainly on two economic models involving Stackelberg differential games in detail. The study of these economic models - one concerned with regional investment allocation problem and the other involving a world industrialization model - form the main contents of this monograph (Chapters 5, 6 and 7). They are preceded by chapters which cover necessary background material on game theory and some of its applications to economics, with special attention given to noncooperative differential games (Chapters 2, 3 and 4). The last chapter (Chapter 8) discusses the largely unexplored terrain of adaptive games. A good background in calculus and elementary statistics is sufficient to follow this book. Although written primarily for systems scientists interested in economic applications, economists and operations researchers will also hopefully find this book useful; particularly, chapters 3,and 4 which provide a short introduction to noncooperative differential games. Both the economic models studied elaborately in this monograph arise in the field of Development Economics.

The research on hierarchical regional investment allocation game reported here was performed by R.C.W. Strijbos, H. Kuilder and myself. World industrialization model was studied under a project of the Central Researchpool of the Twente University of Technology, participated by M. Moraal, G.J. Olsder and myself. My thanks to all of them, and especially to Rens Strijbos, for their contributions to much of the research reported here. My thanks also to Tamer Başar of the University of Illinois, who first enlightened me with various intricate aspects of strategies and nonuniqueness of solutions in games. Huibert Kwakernaak, as usual, was constant source of encouragement and my thanks to him for enthusiastically supporting my idea of writing this monograph. Most of the material contained here was presented in a series of lectures at the Indian Institute of Management, Calcutta, during July-August of 1983. My thanks to the participants of the seminar, and especially to Biswanath Sarkar, who arranged for my visit to the Institute. The manuscript was typed skillfully by Marja Langkamp, Monique IJgosse and in large part, by Manuela Fernández, to all of whom I owe my thanks.

Finally, I am very much indebted to my wife, Arundhati, and my daughter, Mishtu, for their understanding and encouragement during the preparation of this book and in all my intellectual endeavors.

Enschede
March, 1984

Arunabha Bagchi

CONTENTS

CHAPTER 1

INTRODUCTION AND OVERVIEW

1. HISTORICAL BACKGROUND

Game theory is concerned with the mathematical study of conflict and cooperation. Although conflict and cooperation pervade our society through space and time, a systematic study of game theory is a rather recent phenomenon in the history of human thought. Once a theory was developed, the potential application to numerous branches of social science became immediately apparent. Today, game theory plays a central role in modern economic theory and is used as a basic modeling tool in political science, sociology, operations research, military logistics and other fields.

Although attempts have been made earlier to formalize game in mathematical terms, John von Neumann is rightly credited as being the originator of game theory. Twenty years after his proof of the celebrated "minimax theorem", von Neumann published the classic treatise, jointly with Oskar Morgenstern in 1944, entitled Theory of Games and Economic Behavior. The book was rich in many entirely new ideas and possibilities for economic applications. Three different ways of representing a game-in normal form, in extensive form and via characteristic functions - were introduced and several solution concepts were defined, including the so-called von Neumann-Morgenstern solution.
It was only after the publication of this book that one saw an explosive growth of research in game theory. This started in earnest in the early fifties and shows no sign of abetting. Saddle points and equilibrium concepts in normal forms are used in mathematical programming and statistical decision theory. The representation of a game in extensive form enables immediate generalization to multi-act games, stochastic games and also to differential games. Finally, the representation of

games via characteristic functions form the basis for the theory of cooperative games which has a central role in mathematical economics. Major extensions include, among others, Shapley value of a game, games with infinite number of players and Nash's cooperative solution for games without side payments.

Noncooperative games first appeared in the work of A. Cournot as early as 1838 when he modelled a market game in which duopoly prevails. When a market has only a few traders who have influence over the market, one talks about an oligopoly. Duopoly is the special case of two traders. It also appeared in the previously cited treatise of von Neumann and Morgenstern. But the equilibrium solution concept for nonzero-sum N-person noncooperative game in normal form was first systematically studied by Nash (1951). The solution concept introduced by Nash has the troubling feature of being nonunique in many situations. One may sometimes impose some desirable properties of the solution to make the equilibrium solution unique. They are social rationality, strategy dominance and inadmissibility of mixed strategies. In noncooperative games, players typically maximize their respective payoffs (or minimize their respective costs). A strategy tuple is socially rational or Pareto optimal if there is no other strategy with payoffs having the property that at least one player receives more and no player receives less than the payoffs they receive with the strategy tuple under consideration. One strategy tuple dominates another if the payoff resulting from the former associated with each player is at least as large as that resulting from the latter. Mixed strategies refer to the situation when we allow for chance mechanism to determine strategies of the players. Significant progress was made in studying games in extensive form and the role of information there following the publication of the important paper of Kuhn (1953). Stackelberg solution in nonzero-sum games was first introduced by H. von Stackelberg (1934) within the context of economic competition. The concept becomes relevant whenever one (or more) player(s) has (have) dominant role(s) in the game. It is, therefore, a natural concept in hierarchical systems. It also appears in a 2-person game where one player can dictate his will on the other player. Two highly intuitive and original introductory books in game theory are Luce and Raiffa (1957) and Shubik (1975).

From the point of view of optimization, game theory is concerned with many agents interested in optimizing different criteria which may be in conflict with one another. It can be viewed as an extension of standard optimization problems.

Optimization of dynamical systems developed rapidly after the introduction of the idea of dynamic programming by Richard Bellman and the proof of the maximum principle by Pontryagin and his coworkers in the fifties. This field, known as optimal control theory, is involved with determining an optimal control that optimizes a criterion subject to the dynamical constraint expressing the evolution of the system state under the influence of the control term. It is only natural to extend this to the case of multiple controllers (also called decision makers) with different and conflicting optimizing criteria. This is the subject matter of differential games. Zero-sum differential games, also called pursuit-evasion games, was single-handedly created by Isaacs in the early fifties resulting in his famous book (Isaacs 1975; 1st edn. 1965). Research in the sixties concentrated mainly on the rigorous treatment of Isaacs equation (Friedman, 1971). Nonzero-sum differential games were introduced systematically for the first time by Starr and Ho (1969). Informational nonuniqueness of Nash equilibria was studied in a series of papers by Basar (see Basar and Olsder, 1982). There was also significant theoretical advance in the seventies on nonzero-sum deterministic differential games. Stackelberg differential games were first treated in the papers of Chen and Cruz, Jr (1972) and Simaan and Cruz, Jr (1973). Hierarchical (Stackelberg) equilibria and the related theory of "incentives" are active areas of theoretical research at present.

Incentives, a rapidly growing field of research, is not discussed in this book. A lucid tutorial introduction to the subject from a control-theoretic viewpoint may be found in the paper of Ho, Luh and Olsder (1980).

Pontryagin's maximum principle found almost immediate application to economics; specially, to the theory of optimal economic growth (See, for example, K. Shell (ed.) 1967, and Arrow and Kurz, 1970). Stochastic control theory was also applied to economic models (Kendrick, 1981). There is a vast literature on the application of control theory to economics. A reasonably up-to-date review may be found in Pitchford and Turnovsky (1977). Nonzero-sum noncooperative differential games were applied to economic models by Case (1971) for the first time. For recent developments, see Feichtinger, ed. (1982). Some applications of Stackelberg differential games to economic models have appeared in the last few years. For an interesting application to oil embargo, see Hauptmann (1982). Jørgensen (1982) is a recent survey of nonzero-sum differential games applied to advertising models.

2. OUTLINE OF THE BOOK

Chapter 2 gives an overview of game theory and its connection with economics. Thus, it briefly discusses cooperative game in relation to competitive equilibrium in a market game and then studies Nash and Stackelberg equilibrium solution concepts for noncooperative games. Both finite and continuous - kernel games are considered. Economic applications in oligopoly markets and in hierarchical planning models are also discussed. This chapter is primarily intended for systems scientists to give them a short introduction to this vast subject.

In chapter 3, dynamics is brought into the game problem. Nonzero-sum differential games are formulated and equilibrium solution concepts for noncooperative and hierarchical games are explained. The difference between control action and strategy and that between open-loop, feedback and closed-loop strategies are explained . Key results in optimal control theory are then stated without proofs and an economic interpretation of the Pontryagin maximum principle is given. Finally, an application of the maximum principle to the problem of regional allocation of investment is analyzed in reasonable detail. This forms the starting point of Chapter 5.

Chapter 4 gives solutions of Nash and Stackelberg differential games. Linear- quadratic case is also treated extensively. Applications of the results to economics are provided. Thus, dynamic duopoly model is formulated and solved in the set-up of Nash games, while a R & D problem in economics is formulated and solved in the framework of Stackelberg games.

Chapter 5 extends the regional investment allocation problem formulated in chapter 3. First, a dual economy with hierarchical decision making is considered. An element of game is thereby introduced in a two region problem which is studied in detail. Then the problem is formulated in its generality as a N-region investment allocation model with the central planning board as the highest decision making authority. A number of interesting features of this problem are analyzed. Various simulation studies are included for better understanding of the game-theoretic set-up. One major goal of investment allocation is to achieve an "equitable" growth among regions. This necessitates introducing an appropriate measure of regional inequality which is a research area of independent interest. Some basic results in this area are also presented for a complete treatment of the problem.

In chapter 6, a linear programming problem is studied in the context of a leader-follower structure among the decision variables. In the context of the follower, one encounters a standard LP-problem, but the leader has to optimize in a nonconvex region and the problem becomes considerably complicated. In the first place, one may have a multitude of optimal solutions, and in the second place, a suitable algorithm to obtain one of these solutions is far from obvious. One algorithm that appears to work reasonably well is discussed in detail.

Chapter 7 begins with a world industrialization model as developed originally in the Development Planning Department of the Erasmus University at Rotterdam. This model is then aggregated and converted into a Stackelberg linear programming problem considered in the preceding chapter. The nature of the optimal solution obtained for this model is analyzed and some policy implications are discussed.

Finally, in Chapter 8, a related problem area of considerable research potential is introduced. An econometric model is considered where different economic agents play a Stackelberg or Nash difference game and model parameters are unknown. The purpose is to simultaneously estimate the unknown parameters and determine the optimal strategies. This can be termed an adaptive game problem. An interesting suboptimal solution to this problem has recently been proposed. This is explained in detail in this chapter. Systems scientists with research experience in systems identification and adaptive control can contribute substantially to this new area of significant economic applications. So far, this problem has been studied in the context of "rational expectation" models, but the methodology has potential application to the interesting problem where one would like to determine, on the basis of past data, what sort of game has actually been played among the economic agents.

CHAPTER 2

GAME THEORY AND ECONOMICS

1. INTRODUCTION

In this chapter we give a brief introduction to game theory and its application to classical economics. This chapter only serves the role of motivating the reader to appreciate the connection between game theory and economics and is not intended to cover, even partially, this vast subject. Although cooperative games are outside the main scope of this book, we briefly discuss some aspects of them because of their profound importance to the theory of economic equilibrium. We then introduce finite and continuous-kernel (infinite) multi-person nonzero-sum games and the Nash and Stacklberg equilibrium solution concepts. We also apply them to simple economic models. We conclude with the discussion of hierarchical models where the Stackelberg concept of leadership in the decision making process is inherently built into the problem. Standard reference in game theory is Owen (1968).

2. BASIC NOTIONS OF GAME THEORY

In the context of optimization problems, game theory involves the situation of a multiple number of decision makers, called players, with different optimizing criteria. The optimizing criteria are called welfare (payoff) functions (if maximized) or loss (cost) functions (if minimized). A game with N players is called a N-person game. N = 1 degenerates into a usual optimization problem. If all the players in a game can take only a finite number of actions, we call it a finite game. Otherwise, we have an infinite game. If the sum of the criteria of a game is identically zero, we call it a zero-sum game. Otherwise, we have a nonzero-sum game. We shall be concerned exclusively in this book with nonzero-sum games. In a 2-person zero-sum game, the gain (loss) of one player is exactly equal to the loss (gain) of the other player. We then have the situation of total conflict. At the other extreme, we have the case of identical criteria for the players. This is called a team problem. In this situation, the players gain or lose together and it is only natural for them to cooperate totally. We have a cooperative game if the players form coalitions and decide beforehand the mode of playing the game. Otherwise, we have a noncooperative game. We shall be concerned with noncooperative games in this book.

Although cooperative games will not be studied here, we mention briefly the role of cooperative games in studying economic markets. We follow Billera (1979) where one may find a compact but thorough introduction to economic market games. As soon as there are three or more players with the possibility of cooperation, coalition formation becomes feasible. For a N-person game, there are altogether 2^N possible set of coalitions. A cooperative game with side payments allows the payoffs to be transferred between the players. The theory of such games dates back to the publication of the classic treatise of von Neumann and Morgenstern (1947). Since side payments are allowed, only the total payoff to each possible coalition is of interest. By a N-person cooperative game with side payments (in the characteristic function form), we mean a function

$$v : 2^{\{1,2,\dots,N\}} \longrightarrow \mathbb{R} .$$

For an up-to-date treatment of cooperative games, see Vorob'ev (1977).

By the core of a game v, we mean the set of allocations of $v(\{1,2,\dots,N\})$ which cannot be objected to by any coalition. Thus, if x_n of the total payoff is allocated to player n; that is $\sum_{n=1}^{N} x_n = v(\{1,2,\dots,N\})$, the N-tuple $(x_1,x_2,\dots,x_N)$ belongs to the core of v if, for any coalition $S \subset \{1,2,\dots,N\}$, $\sum_{n \in S} x_n \geq v(S)$. The core gives an important characterization of competitive equilibria in a market economy.

Consider a market with m commodities and N traders. Since side payments are allowed, it must be possible to price each commodity via a common unit so that each trader is able to measure his payoff in terms of that unit. We, therefore, introduce (m+1)-st commodity, money, into the model. If the n-th trader has a commodity vector x^n (of m commodities) available to him, he is able to convert it to a utility $u_n(x^n)$, where the utility function $u_n : \mathbb{R}^m_+ \longrightarrow \mathbb{R}$ ($\mathbb{R}^m_+$ the positive orthand in $\mathbb{R}^m$) is usually assumed to be concave and continuous. Suppose that $w^n \varepsilon \mathbb{R}^m_+$ is the initial commodity vector in possession of the n-th trader. Suppose that a coalition $S \subset \{1,2,...,N\}$ decides to trade together. By a <u>feasible S-allocation</u>, we mean a collection x^n, $n \varepsilon S$, of commodity vectors such that $x^n \geq 0$ (component wise) and $\sum_{n\varepsilon S} x^n = \sum_{n\varepsilon S} w^n$. We now define a <u>market game</u> v by

$$v(S) = \max. \sum_{n\varepsilon S} u_n(x^n) \tag{2.1}$$

where the maximum is taken over all feasible S-allocations. Suppose that the i-th commodity has price p_i (with the price of money taken to be unity). Trader n is interested in an allocation (x^n, ξ^n) of goods and money which maximizes his payoff (utility):

$$U_n(x^n, \xi^n) = u_n(x^n) + \xi^n \tag{2.2}$$

subject to the budget constraint

$$\sum_{i=1}^{n} p_i \; x_i^n + \xi^n = \sum_{i=1}^{m} p_i \; w_i^n \tag{2.3}$$

and with $\xi^n \varepsilon \mathbb{R}$.

The price vector $p = (p_1,...,p_m)'$ is said to be in <u>competitive equilibrium</u> if the optimal holding (x^n, ξ^n) of trader n, $n = 1,...,N$, acting independently to maximize their payoffs, satisfy $\sum_{n=1}^{N} x^n = \sum_{n=1}^{N} w^n$ and $\sum_{n=1}^{N} \xi^n = 0$. Clearly, maximizing $U_n(x^n, \xi^n)$ is equivalent to maximizing $u_n(x^n) - \sum_{i=1}^{m} p_i x_i^n$. Suppose that z^n is a feasible $\{1,2,...,N\}$- allocation that maximizes the latter expression over all $x^n \varepsilon \mathbb{R}^m_+$. (p, z^n), $n = 1,...,N$, is called a competitive solution for the market and a necessary and sufficient condition for this to hold is that the price vector p is in competitive equilibrium. Now the competitive payoff for a market game v is a vector $\alpha \in \mathbb{R}^N$ defined by

$$\alpha_n = u_n(z^n) - \sum_{i=1}^{m} p_i(z_i^n - w_i^n) \qquad (2.4)$$

with (p, z^n), $n = 1,\ldots,N$, a competitive solution for the market. The key result that ties core with competitive equilibria is: "any competitive payoff for a market game is in the core". The existence of competitive solution for a market has been one central area of research for mathematical economists during the fifties and sixties and one can prove, under fairly general conditions, that a competitive solution of a market always exists (and therefore, the market game has a nonempty core).

One can also consider games without side payments. This situation arises, for example, when we rule out the possibility of a cardinal measure of utility (in terms of money) and do not consider money as a commodity anymore. In this case, the function v has to be replaced by a correspondence (set-valued mapping) and the theory becomes more complicated. We do not pursue this here any further.

Noncooperative games arise when the possibility of coalition formation is ruled out. Such games are described either in the normal form or in the extensive form. A game is described in the normal form if all the possible actions or strategies of all the players are specified and the payoffs of each of the players resulting from different combinations of strategies chosen by the players are explicitly given. In a N-player finite game, suppose that the k-th player Pk has available strategies $s_1^k,\ldots,s_{m_k}^k$. If Pk chooses strategy $s_{n_k}^k$, $1 \le n_k \le m_k$, the payoff of the k-th player Pk is given by $J^k(s_{n_1}^1,\ldots,s_{n_N}^N)$, $k = 1,\ldots,N$. A 2-person game is then completely characterized by a pair of matrices A and B with the components given by

$$a_{ij} = J^1(s_i^1, s_j^2) \text{ and } b_{ij} = J^2(s_i^1, s_j^2) \qquad (2.5)$$

These games are, therefore, called bimatrix games. For a zero-sum game, $b_{ij} = -a_{ij}$, the game is characterized by one matrix and is called matrix game. Going back to the general N-person game, we note that there is a one-to-one correspondence between $s^k_{n_k}$ and n_k. Writing $J^k_{n_1,\ldots,n_N}$ for $J^k(s^1_{n_1},\ldots,s^N_{n_N})$, we can, therefore, equivalently refer to n_k as a strategy for Pk, $1 \le n_k \le m_k$ and $J^k_{n_1},\ldots,_{n_N}$ as the payoff for the player Pk.

One drawback of the normal form description is that it gives no information about the way the game is being played. The order of play, information available to the players at different stages and the evolution of the game in the dynamic case, are not recognizable in the normal form. The extensive form specifies all these in detail and is expressed in the form of a tree structure. The tree contains various nodes and branches which provide explicit description of the order of play and the information available to each player at the time of his making decision. The game starts at the vertex and ends at the terminal node of one of the branches. The information available to a player is indicated by dotted lines enclosing an area containing the relevant nodes. We illustrate this by means of two simple diagrams:

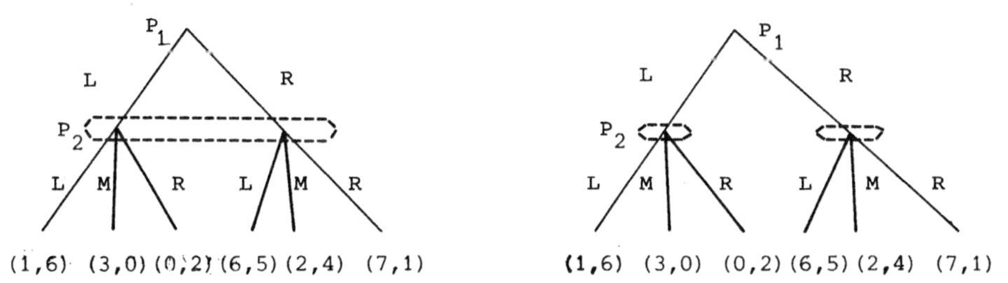

Fig. 2.1 Two nonzero-sum games in extensive forms differing only in the information structure.

Strategies available to P1 are L and R, and to P2 are L, M and R in a 2-person game. In fig. 2.1(a), the two possible nodes of P2 are enclosed by dotted lines. This means that P2 does not know P1's action. It reflects either the situation when both the players act simultaneously or, when P1 acts earlier but P2 has no knowledge of P1's action. We can easily convert this to the normal form of a bimatrix game:

A (P_1 rows, P_2 columns):

	L	M	R	
L	1	3	0	1
R	6	2	7	2
	1	2	3	

B (P_1 rows, P_2 columns):

	L	M	R	
L	6	0	2	1
R	5	4	1	2
	1	2	3	

In fig. 2.1(b), each node of P2 is enclosed in a separate information set which means that P1 acts first and P2 knows exactly P1's action. We can also convert this to a normal form, but interestingly enough, the resulting bimatrix game will be different from the previous one. So far, we have used action and strategy synonymously. It is at this point that we have to make a distinction between them, As before, P1 has two choices L and R. At the outset, P2 makes a strategy for his action which tells him how to react to P1's action. A strategy for P2 is, therefore, a mapping from the set of actions of P1 to that of P2. Denote this mapping by γ^2. P2 has now $3^2 = 9$ possible strategies to choose from (with u^1 denoting the action of P1):

$$\gamma_1^2(u^1) = u^1, \quad \gamma_2^2(u^1) = L, \quad \gamma_3^2(u^1) = M, \quad \gamma_4^2(u^1) = R,$$

$$\gamma_5^2(u^1) = \begin{cases} L & \text{if } u^1 = L \\ M & \text{if } u^1 = R \end{cases} \qquad \gamma_6^2(u^1) = \begin{cases} R & \text{if } u^1 = L \\ L & \text{if } u^1 = R \end{cases}$$

$$\gamma_7^2(u^1) = \begin{cases} M & \text{if } u^1 = L \\ R & \text{if } u^1 = R \end{cases} \qquad \gamma_8^2(u^1) = \begin{cases} M & \text{if } u^1 = L \\ L & \text{if } u^1 = R \end{cases}$$

$$\gamma_9^2(u^1) = \begin{cases} R & \text{if } u^1 = L \\ M & \text{if } u^1 = R \end{cases} \tag{2.6}$$

Note that there is still no conceptual difference between action and strategy for P1. For a given action of P1, the choice of action P2 is based on one of his 9 strategies. Thus, we have now two 2 × 9 matrices of payoffs in the equivalent normal form description of the game as illustrated below:

A (P_1 rows, P_2 columns)

P_1 \ P_2	1	2	3	4	5	6	7	8	9
1	1	1	3	0	1	0	3	3	0
2	7	6	2	7	2	6	7	6	2

B (P_1 rows, P_2 columns)

P_1 \ P_2	1	2	3	4	5	6	7	8	9
1	6	6	0	2	6	2	0	0	2
2	1	5	4	1	4	5	1	5	4

We have considered here only pure strategies, meaning that the mapping γ's are deterministic functions. We can extend this to the case of mixed strategies where a prescribed probability law is used to "mix" the pure strategies during the play of the game. Many other extensions and specifications are possible. Thus, one may have behaviorial strategies, perfect or imperfect information, perfect or imperfect recall, dynamic or multistage games etc. For a short introduction to all these concepts, see Weber (1979).

3. N-PERSON NONCOOPERATIVE GAMES

We consider only static games here. Dynamic games will be introduced in the next chapter. We start with finite games in normal form. Since each player optimizes his own criterion, there is no unambiguous way of defining a solution to this problem. Two solution concepts will be introduced in this section both of which exhibit a sort of equilibrium behavior and are, therefore, also called equilibrium solutions.

Suppose that there is no cooperation among the players and they make decisions independently. One natural solution is a N-tuple of strategies where there is no incentive for any player to deviate unilaterally from that point.

Definition 2.1

A N-tuple of strategies $\{n_1^*,\ldots,n_N^*\}$ is called a Nash equilibrium solution for a N-person noncooperative static finite game formulated in the previous section if

$$\begin{aligned} J^{1^*} &\triangleq J^1_{n_1^*,n_2^*\ldots,n_N^*} \geq J^1_{n_1,n_2^*,\ldots,n_N^*} \\ &\vdots \\ J^{N^*} &\triangleq J^N_{n_1^*,n_2^*,\ldots,n_N^*} \geq J^N_{n_1^*,n_2^*,\ldots,n_N} \end{aligned} \tag{2.7}$$

Here the N-tuple $(J^{1^*};\ldots,J^{N^*})$ is called a Nash equilibrium outcome of the game.

In the example corresponding to fig. 2.1(a), it is obvious that (2,1) is the unique Nash equilibrium solution with (6,5) being the Nash equilibrium outcome of the game. One can easily construct examples to show that, in the class of pure strategies, a Nash equilibrium solution need not exist, and if it exists, it does not have to be unique. We would naturally like to order different Nash equilibrium solutions among themselves so as to determine the best one for our purposes. One can narrow down the Nash equilibrium solutions to the so-called "admissible" one. A strategy $\{n_1,\ldots,n_N\}$ is said to be "better" than another strategy $\{l_1,\ldots,l_N\}$ (strategy dominance) if

$$J^i_{n_1,\ldots,n_N} \geq J^i_{l_1,\ldots,l_N} \quad \text{for all } i = 1,\ldots,N \tag{2.8}$$

A Nash equilibrium solution is said to be admissible if there exists no better Nash equilibrium solution. It is, however, easy to find simple examples to show that there may be more than one admissible Nash equilibrium solution, the components of which are not interchangeable and this leads to the possibility of the outcome of the game to be a nonequilibrium one in the actual play. Finally, we mention that if we allow for mixed strategies, the noncooperative game formulated above has always at least one Nash equilibrium solution. This important result is due to Nash (1951). The corresponding celebrated result for 2-person zero-sum games, known as the minimax theorem, was originally proved by von Neumann.

There is another equilibrium solution concept in nonzero-sum games which becomes relevant when one or several players has (have) dominant role(s) vis-a-vis the rest of the players. One has to introduce in this situation a hierarchical equilibrium concept which was first suggested by von Stackelberg (1934) and is, therefore, referred to as the Stackelberg equilibrium solution concept. The definition becomes extremely messy for N-person games and for clarifying the essential ideas, we consider only 2-person games here. We write M_k to denote the set $\{1,2,\ldots,m_k\}$. The player holding the dominant position in the game is called the leader, while the other player is called the follower. Suppose that P1 is the leader who can impose his decision on P2. Thus, if P1 chooses a strategy $n_1 \in M_1$, P2 has to maximize his criterion as a reaction to the strategy n_1 announced by P1. Suppose that there is a unique $n_2 \in M_2$ that maximizes $J^2_{n_1,1}$ for $1 \in M_2$. Then n_2 is a function of n_1, say, $t(n_1)$. The optimal strategy for P1 is now obvious. A strategy n_1^* is optimal for P1 if

$$J^1_{n_1^*,t(n_1^*)} \geq J^1_{n_1,t(n_1)} \quad \forall\, n_1 \in M_1 \tag{2.9}$$

The following definition is formalization and generalization of the preceding intuitive discussion.

Definition 2.2

In a bimatrix game with P1 as the leader, the optimal reaction set $R^2_{n_1}$ of P2 to the strategy $n_1 \in M_1$ of P1 is defined by

$$R^2_{n_1} = \{1 \in M_2 \,|\, J^2_{n_1,1} \geq J^2_{n_1,n_2} \;\forall\, n_2 \in M_2\} \tag{2.10}$$

The uniqueness assumption above means that $R^2_{n_1}$ is a singleton for all n_1. In a 2-person finite game with P1 as the leader, a strategy $n_1^* \in M_1$ is called a Stackelberg equilibrium solution for the leader if

$$\min_{n_2 \in R^2_{n_1^*}} J^1_{n_1^*,n_2} \geq \min_{n_2 \in R^2_{n_1}} J^1_{n_1,n_2} \qquad \forall\, n_1 \in M_1 \tag{2.11}$$

If $n_1^* \in M_1$ is a Stackelberg solution for the leader P1, any element $n_2^* \in R^2_{n_1^*}$ is an optimal strategy for the follower P2 that is in equilibrium with n_1^*. The pair $\{n_1^*,n_2^*\}$ is a Stackelberg equilibrium solution with P1 as the leader and $\{J^1_{n_1^*,n_2^*}, J^2_{n_1^*,n_2^*}\}$ is the corresponding Stackelberg equilibrium outcome of the game.

Let us illustrate this by means of the example corresponding to fig. 2.1(a). Suppose that P1 is the leader. It is easy to see that $R^2_1 = \{1\}$ and $R^2_2 = \{1\}$ are both singletons, Furthermore, $J^1_{2,1} = 6 > J^1_{1,1} = 1$ implying that (2,1) is the unique Stackelberg equilibrium solution, just as it is the unique Nash equilibrium solution, for the game. On the other hand, if the matrix B of the bimatrix game considered here is altered slightly by interchanging the (2,1)th and (2,2)th elements, the resulting game problem has no Nash equilibrium solution in the class of pure strategies, while (2,2) is the unique Stackelberg solution with P1 as the leader. It can be easily proved that every 2-person finite game admits a Stackelberg equilibrium solution in the class of pure strategies. The equilibrium solution does not have to be unique. However, the Stackelberg payoff for the leader, $J^1_{n_1^*,n_2^*}$ in the definition 2.2, is unique and we do not encounter the same problem that arises due to the nonuniqueness of Nash equilibrium solution.

Let us now consider static infinite games; that is, games where at least one of the players has, at its disposal, an infinite number of actions or strategies. If the action space for each player is a continuum and the cost functional of each player is continuous in all of its arguments, we talk of continuous-kernel games. It is rather straightforward to introduce Nash and Stackelberg equilibrium solution concepts for such games, in analogy with our preceding definitions 2.1 and 2.2. We have to replace max. and min. appearing in definition 2.2 by sup. and inf. respectively. We consider 2-person continuous-kernel games to give geometric interpretation of Nash and Stackelberg equilibrium solutions. Suppose that $J^i(u^1,u^2)$ is the payoff function for Pi, i = 1,2, with $u^i \in U^i$. Clearly, the set

$$R^2(u^1) = \{\xi \in U^2 | J^2(u^1,\xi) \geq J^2(u^1,u^2), \ \forall \ u^2 \in U^2\}$$

is the optimal reaction set for P2. If $R^2(u^1)$ is a singleton for every $u^1 \in U^1$, it is called the reaction curve for P2. For the graphical illustration below, we follow Simann and Cruz, Jr. (1973). The payoff functions are replaced by cost functions and we are interested in minimizing J^1 and J^2 instead of maximizing them. This means, of course, that working with $-J^1$ and $-J^2$ would agree with our original problem formulation. We assume that $U^1 = U^2 = \mathbb{R}$ and both J^1 and J^2 are convex and twice differentiable with respect to u^1 and u^2, having level curves shown in the figure 2.2 below. The reaction sets are then singletons so that we can talk of reaction curves. These reaction curves, 1^1 and 1^2, are obtained by joining the points of tangency between the level curves and the lines of constant u^2 and u^1, respectively. By definition, the Nash solution must lie on both reaction curves and if these curves have one point of intersection (as in figure 2.2), the Nash equilibrium solution for the game exists and is unique. Now consider the Stackelberg game with P1 as the leader. Then P1 seeks for a minimum over 1^2 and this is achieved at a point where the level curve of J^1 is tangent to 1^2. On the other hand, if P2 is the leader, the Stackelberg equilibrium solution is attained at a point where the level curve of J^2 is tangent to the reaction curve 1^1. In figure 2.2 below, N is the Nash solution while S^1 is the Stackelberg solution with P1 as the leader and S^2 is the Stackelberg solution with P2 as the leader. In this particular example, the Nash equilibrium costs of both players are higher than their corresponding Stackelberg equilibrium costs. This, however, need not be true always. One can easily visualize situations where the follower is "worse off" in the "Stackelberg game" than in the corresponding "Nash game". The assertion that always holds, in the case that $R^2(u^1)$ is a singleton for all $u^1 \in U^1$, is that the leader cannot be "worse off" in the "Stackelberg game" than in the corresponding "Nash game". Finally, figure 2.3 illustrates a slightly modified game where Stackelberg equilibrium solution exists while a Nash equilibrium solution does not exist.

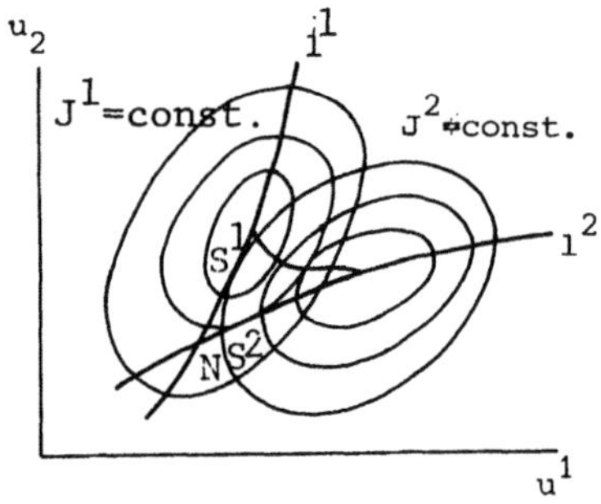

Fig. 2.2 Graphical illustration of a game with Nash and Stackelberg solutions

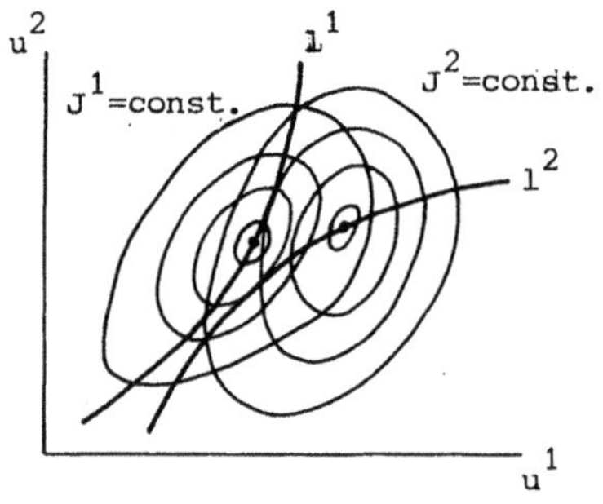

Fig. 2.3 Graphical illustration of a game without Nash but with Stackelberg solutions.

Continuous kernel games with quadratic cost functions have attracted widespread attention because of their inherent simplicity and analytic tractability. We give below explicit Nash and Stackelberg equilibrium solutions for this class of games, called quadratic games, for the special case of N = 2. Suppose that

$$J^i = \frac{1}{2}\,(u^{1'}R^i_{11}u^1+u^{1'}R^i_{12}u^2+u^{2'}R^i_{21}u^1+u^{2'}R^i_{22}u^2) + r^{i'}_1 u^1+r^{i'}_2 u^2+c^i \quad , \; i = 1,2 \tag{2.12}$$

with $u^i \in U^i = \mathbb{R}^{m_i}$, $i = 1,2$. R^i_{jk} is a $(m_j \times m_k)$-matrix for $j,k = 1,2$ with $R^i_{ii} > 0$, $i = 1,2$; r^i_1 and r^i_2 are m_1 and m_2-dimensional vectors respectively while c^i is a constant, for $i = 1,2$. There is no loss of generality in assuming that $R^{i'}_{12} = R^i_{21}$ and that R^i_{22} is a symmetric matrix.

Theorem 2.1

The 2-person quadratic game defined by cost function (2.12) admits a unique Nash equilibrium solution if and only if the following equation

$$Ru = -r \tag{2.13}$$

where

$$R = \begin{bmatrix} R^1_{11} & R^1_{12} \\ R^2_{12} & R^2_{22} \end{bmatrix} , \quad u' = (u^1,u^2), \quad r' = (r^1_1 \; r^2_2)$$

admits a solution. This Nash solution is unique if the matrix R defined above is invertible, in which case this unique Nash solution is given by

$$u_N = -R^{-1}r \tag{2.14}$$

Proof To obtain the Nash equilibrium solution, we differentiate J^i with respect to u^i, $i = 1,2$, set the resulting expressions equal to zero and solve the pair of equations thus obtained. Because of strict convexity of our cost functions, this also provides a sufficient condition for the Nash equilibrium solution. Differentiating J^i with respect to u^i yields

$$R^i_{ii}u^i + R^i_{ij}u^j + r^i_i = 0, \; i = 1,2, \; j \neq i \tag{2.15}$$

which can be rewritten as

$$Ru = -r$$

The rest follows trivially.

Remark

This result can be easily extended to the case of N-person nonzero-sum quadratic games.

Let us now turn to the Stackelberg equilibrium solution for this game with P1 as the leader. The following theorem gives explicit expression for the solution:

Theorem 2.2

Suppose that the matrix

$$R \equiv R^1_{11} + R^{2'}_{21}(R^2_{22})^{-1}R^1_{22}(R^2_{22})^{-1}R^2_{21} - R^1_{21}(R^2_{22})^{-1}R^2_{21} - R^{2'}_{21}(R^2_{22})^{-1}R^{1'}_{21} > 0 \tag{2.16}$$

Then our 2-person game problem admits a unique Stackelberg equilibrium solution with P1 as the leader, given by

$$u^1_{S1} \equiv -R^{-1}[R^{2'}_{21}(R^2_{22})^{-1}R^1_{22}(R^2_{22})^{-1}r^2_2 - R^1_{21}(R^2_{22})^{-1}r^2_2 + \\ + r^1_1 - R^{2'}_{21}(R^2_{22})^{-1}r^1_2] \tag{2.17a}$$

$$u^2_{S1} = -(R^2_{22})^{-1}[R^2_{21}u^1_{S1} + r^2_2] \tag{2.17b}$$

Proof Given u^1, the follower minimizes u^2 and the unique minimum is given, from eq^n. (2.15), by

$$R^2_{22}u^2 + R^2_{21}u^1 + r^2_2 = 0 \tag{2.18}$$

To obtain the Stackelberg equilibrium solution for the leader, we have to minimize J^1 with respect to u^1 for $u^1 \in \mathbb{R}$, subject to the constraint (2.17). From (2.17), we can write u^2 uniquely in terms of u^1(R^2_{22} is invertible, by assumption). Substituting this in J^1, we can express the criterion solely as a function of u^1:

$$\tilde{J}^1(u^1) = \tfrac{1}{2}u^{1'}R^1_{11}u^1 + \tfrac{1}{2}[R^2_{21}u^1+r^2_2]'(R^2_{22})^{-1}R^1_{22}(R^2_{22})^{-1}$$
$$[R^2_{21}u^1+r^2_2] - u^{1'}(R^2_{22})^{-1}[R^2_{21}u^1+r^2_2]+ u^{1'}r^1_1$$
$$-[R^2_{21}u^1+r^2_2]'\ (R^2_{22})^{-1}r^1_2 + c^1.$$

Assuming that R given by (2.16), $\tilde{J}^1(u^1)$ has a unique minimum with respect to u^1 for $u^1 \in \mathbb{R}$ and the minimum is attained for u^1_{S1} given by (2.17a). It follows immediately from (2.18) that the corresponding unique strategy, u^2_{S1}, of P2 is given by (2.17b). □

4. ECONOMIC APPLICATIONS

The standard application of static Nash game to economics is the study of oligopolistic markets. In an oligopoly, buyers cannot influence the price and quantities offered by the traders and the collective behavior of the buyers is assumed to be fixed and known. Two standard sources of application of Nash games to economics are Friedman (1977) and Case (1979). The latter book in highly original and points out the crucial importance of "robustness" aspect of the equilibrium solution for applying noncooperative game theory to economic problems.

Example 2.1 (Cournot Duopoly Model)

Consider two firms F1 and F2 producing identical products. Let q^1 and q^2 be the outputs of the two firms and p is the price of the product. Clearly, p is related to the total output $q^1 + q^2$ by means of a demand function. Suppose that the demand function is known and is differentiable. Thus

$$p = h(q^1+q^2) \qquad (2.19)$$

It is intuitively clear that $h(\cdot)$ should be a monotonically decreasing function of its argument, since an increase in total supply will cause a decrease in the price of the product.

Suppose that the production cost functions of the two firms are given by $c^i(q^i)$, $i = 1,2$, so that the profit of the firm Fi that it wants to maximize is given by

$$\Pi^i(q^1,q^2) = q^i h(q^1+q^2) - c^i(q^i); \; i = 1,2 \tag{2.20}$$

It is usual to assume $c^i(\cdot)$ to be a convex function of its argument, $i = 1,2$. The outputs q^1 and q^2 of the two firms are at their disposal. They are free to decide on their respective outputs so as to maximize their individual profits. If the firms act rationally and have no collusion between them, the Nash equilibrium solution becomes appropriat The nature of the Nash equilibrium concept means that the production levels set by the two firms are decided independently; that is, there is no dependence of q^1 on q^2 and vice versa. The optimality conditions for Nash equilibrium solution are given by

$$\frac{\partial \Pi^1}{\partial q^1} = h(q^1+q^2) - \frac{dc^1(q^1)}{dq^1} = 0 \tag{2.21a}$$

$$\frac{\partial \Pi^2}{\partial q^2} = h(q^1+q^2) - \frac{dc^2(q^2)}{dq^2} = 0 \tag{2.21b}$$

under the assumption that $q^1, q^2 \in \mathbb{R}$. These equations have to be solved to determine an equilibrium pair (q^{1*},q^{2*}).

If q^1 did depend on q^2 and/or q^2 on q^1, optimality conditions for equilibrium solution would be given by

$$\frac{d\Pi^1}{dq^1} = h(q^1+q^2) + q^1 \frac{\partial h}{\partial q^1} + q^1 \frac{\partial h}{\partial q^2}\frac{dq^2}{dq^1} - \frac{dc^1(q^1)}{dq^1} = 0 \tag{2.22a}$$

$$\frac{d\Pi^2}{dq^2} = h(q^1+q^2) + q^2 \frac{\partial h}{\partial q^2} + q^2 \frac{\partial h}{\partial q^1}\frac{dq^1}{dq^2} - \frac{dc^2(q^2)}{dq^2} = 0 \tag{2.22b}$$

The terms $\frac{dq^2}{dq^1}$ and $\frac{dq^1}{dq^2}$ are called the <u>conjectural variations</u> terms in economics and they reflect the effect of variations in the output of one firm on the output of the other one. We shall come back to these equations later.

Going back to the Nash game, let us assume that

$h(x) = a-bx$, a and b positive constants

and

$c^i(x) = k_i x^2$, k_i nonnegative constants, $i = 1,2$.

Then equations (2.21-ab) can be solved to yield

$$q^{1*} = \frac{a(b+2k_2)}{3b^2+4k_1k_2+4b(k_1+k_2)} \quad ; \quad q^{2*} = \frac{a(b+2k_1)}{3b^2+4k_1k_2+4b(k_1+k_2)} \tag{2.23}$$

In the trivial case of no production cost (two sellers selling water to villagers from their wells), $k_1 = k_2 = 0$ and we get

$$q^{1*} = q^{2*} = \frac{a}{3b} ,$$

and consequently,

$$\pi^{1*} = \pi^{2*} = \frac{a^2}{9b} .$$

Suppose now that a Stackelberg game is being played by the firms with F1 acting as the leader. Then the optimal production level q^2 of F2 will depend on the production level q^1 of F1. We can obtain the Stackelberg equilibrium solution by solving equations (2.22a-b) where $\frac{dq^1}{dq^2} = 0$ and $\frac{dq^2}{dq^1}$ is evaluated from the relation $q^2 = (a-bq^1)/[2(b+k_2)]$, which gives the reaction curve of F2. We can also obtain the equilibrium solution by using theorem 2.2. In that notation,

$$R^1_{11} = 2(b+k_1),\ R^1_{12} = R^1_{21} = b,\ R^1_{22} = 0,\ r^1_1 = -a,\ r^1_2 = 0,\ c^1 = 0;$$

$$R^2_{11} = 0,\ R^2_{12} = R^2_{21} = b,\ R^2_{22} = 2(b+k_2),\ r^2_1 = 0,\ r^2_2 = -a,\ c^2 = 0.$$

Eq^n. (2.17a-b) gives us

$$q^1_{S1} = \frac{ab+2ak_2}{2(b^2+2b(k_1+k_2)+2k_1k_2)} \tag{2.24a}$$

and

$$q^2_{S1} = \frac{1}{2(b+k_2)} (a-bq^1_{S1}). \tag{2.24b}$$

In the special case of no production cost ($k_1=k_2=0$),

$$q^1_{S1} = \frac{a}{2b} \text{ and } q^2_{S1} = \frac{a}{4b},$$

and consequently,

$$\pi^1_{S1} = \frac{a^2}{8b} \text{ and } \pi^2_{S1} = \frac{a^2}{16b}.$$

Comparing with the corresponding result for Nash game, we see that F1 is better off when acting as a leader. There is always advantage in being a leader!

Example 2.2 (Generalization to Oligopolistic Market)

We generalize the Cournot duopoly game to the oligopolistic situation, allowing for, in addition, product differentiation and nonsymmetry with any number of players. After studying the Cournot game, where the quantities produced by the firms are the decision variables, we also consider the Edgeworth game, where the firms use prices as their strategies. They are called "quantity game" and "price game", respectively. The details may be found in Leviatan and Shubik (1971), where the results are also applied to the automobile market in the United States.

We use the following notations:

p^i : price charged by firm i, $i = 1,\ldots,n$

π^i : profit of the firm i

w^i : weighting factor reflecting the nonsymmetry in the market of the product of the firm i

$\sum_{i=1}^{n} w^i = 1$

$\bar{p}$: weighted average price $= \sum_{i=1}^{n} w^i p^i$

β : price sensitivity of overall demand

V : price at which demand is zero

c^i : unit average cost of the firm i

K^i : fixed costs for the firm i.

q^i : quantity produced by the firm i.

In the case of undifferentiated product, price is easily determined by the demand function (as in example 2.1):

$$p = h(\sum_{i=1}^{n} q^i).$$

The situation is complicated in the case of differentiated products. Suppose that d^i is the demand for the product of Fi at the price level p^i. Assume the equilibrium situation when the market is cleared; that is, $d^i = q^i$. Suppose that d^i is given by means of the relation:

$$d^i = \beta w^i (V - p^i - \gamma(p^i - \bar{p})), \quad \gamma > 0 \text{ a constant} \tag{2.25}$$

The profit for Fi can be expressed as

$$\Pi^i = (p^i - c^i)(\beta w^i (V - p^i - \gamma(p^i - \bar{p})) - K^i \tag{2.26}$$

We use the following notations:

S a $n \times n$ square matrix with ij-th element equal to 1 for all $i,j = 1,\ldots,n$

$\hat{1}$ a $n \times 1$ vector with i-th element equal to 1 for all $i = 1,\ldots,n$.

$W = \text{diag.}\ (w^1,\ldots,w^n)$

$q = (q^1,\ldots,q^n)'$

$p = (p^1,\ldots,p^n)'$

$c = (c^1,\ldots,c^n)'$

I denotes the identity matrix.

Equating d^i with q^i, we can write eq^n. (2.25) as

$$q = \beta W(V\hat{1} - ((1+\gamma)I - \gamma SW)p) \tag{2.27}$$

which can be solved for p to yield

$$p = V\hat{1} - \frac{1}{\beta(1+\gamma)} (W^{-1}+\gamma S)q \tag{2.28}$$

The profit (payoff) of Fi can then be expressed as

$$\Pi^i = q^i(V - \frac{1}{\beta(1+\gamma)} (\frac{q^i}{w^i} + \gamma\sum_j q^j) - c^i) - K^i \tag{2.29}$$

To obtain a Nash equilibrium solution, we set $\partial\Pi^i / \partial q^i = 0$, $i = 1,\ldots,n$, and solve this set of equations. In matrix notation, this can be expressed as:

$$\frac{\gamma}{\beta(1+\gamma)} (\frac{1}{\gamma}(2W^{-1}+\gamma I)+S)q = V\hat{1} - c \tag{2.30}$$

Using the identity

$$(Z^{-1}+S)^{-1} = Z - \rho ZSZ$$

where $Z = \text{diag.}\ (z^1,\ldots,z^n)$

$$z^i = \frac{\gamma w^i}{2+\gamma w^i}$$

and

$$\rho = \frac{1}{1+\sum_i z^i} \quad ,$$

eq^n. (2.30) can be solved to yield

$$q = \frac{\beta(1+\gamma)}{\gamma} (Z-\rho ZSZ)(V\hat{1}-c) \tag{2.31}$$

and substituting this in eq^n. (2.28) gives

$$p = V\hat{1} - \frac{1}{\gamma}(W^{-1}+\gamma S)(Z-\rho ZSZ)(V\hat{1}-c) \tag{2.32}$$

In Levitan and Shubik (1971), eq^n. (2.32) was further simplified to get, for $i = 1,\ldots,n$,

$$p^i = \rho\ (V+\gamma\sum_j \frac{w^j c^j}{2+\gamma w^j})\ \{\frac{1+\gamma w^i}{2+\gamma w^i}\} + \{\frac{c^i}{2+\gamma w^i}\} \tag{2.32'}$$

In the symmetric case when

$$w^i = \frac{1}{n} \quad \text{for all } i$$

and

$$c^i = \bar{c} \quad \text{for all } i,$$

$$p^i = \frac{n+\gamma}{2n+(n+1)\gamma}\left(V+\frac{\gamma n}{2n+\gamma}\bar{c}\right)+\frac{n}{2n+\gamma}\bar{c}. \qquad (2.33)$$

Let us now turn to the price game where the price p^i is the decision variable for Fi. We see that the profit (payoff) of Fi given by eq^n. (2.26) is already expressed as function of $p^1,\ldots,p^n$. For a noncooperative Nash equilibrium, we must solve the set of equations obtained by setting the derivatives of Π^i given by (2.26) with respect to p^i equal to zero:

$$\frac{\partial \Pi^i}{\partial p^i} = \beta w^i[-(2(1+\gamma)-\gamma w^i)p^i + \gamma\sum w^j p^j + V + \{(1+\gamma(1-w^i)\}c^i] = 0 ;$$

or in matrix notation,

$$(\Delta I-W-SW)p = \frac{V}{\gamma}\hat{1} + (\tfrac{\Delta}{2}I-W)c \qquad (2.34)$$

where

$$\Delta = \frac{2(1+\gamma)}{\gamma}.$$

We have to invert $(\Delta I-W-SW)$ to determine the optimal p. We write this matrix equivalently as $(X-S)W$, where $X = \Delta W^{-1} - I$. Let $(X-S)^{-1} = Y$. The ij-th component y_{ij} of Y can be expressed as

$$y_{ij} = \frac{R_j+\delta_{ij}}{x_i} = \frac{R_i+\delta_{ji}}{x_j} \quad \text{(because of symmetry)}$$

where $R_j = \sum_k y_{k_j}$ and $X = \text{diag.}(x_1,\ldots,x_n)$. This implies, for $i \neq j$,

$$\frac{R_j}{x_i} = \frac{R_i}{x_j} \quad ; \quad \text{i.e., } R_i x_i = Q, \text{ a constant.}$$

Then we have

$$y_{ij} = \begin{cases} \dfrac{Q}{x_i x_j} & \text{for } i \neq j \\ \dfrac{Q}{x_i^2} + \dfrac{1}{x_i} & \text{for } i = j \end{cases}$$

It follows then that

$$(X-S)^{-1} = QX^{-1}SX^{-1} + X^{-1}$$

and the equality

$$\frac{Q}{x_i} = R_i = \sum_j y_{ij} = \sum_j \frac{Q}{x_i x_j} + \frac{1}{x_i}$$

gives us

$$Q = Q \sum_j \frac{1}{x_j} + 1 = \frac{1}{1-\sum_j \frac{1}{x_j}} .$$

We then have

$$p = W^{-1}(QX^{-1}SX^{-1}+X^{-1})(\frac{V}{\gamma}\hat{1}+(\frac{\Delta}{2}I-W)c). \quad (2.36)$$

Using the identities

$\frac{1}{x_i} = w^i/_{\Delta-w^i}$ and $\sum_i \frac{1}{x_i} = 1 - 1/Q$, we can rewrite

eq^n. (2.5) as

$$p^i = \frac{1}{\Delta-w^i} \{\frac{V}{\gamma}Q+\frac{1}{2} [(\Delta-2w^i)c^i + Q\sum_j\frac{1}{x_j} (\Delta-2w^j)c^j]\} \quad (2.36')$$

In the symmetric case when $w^i = \frac{1}{n}$ and $c^i = \bar{c}$ for all i, we obtain

$$p^i = \frac{n}{2n+(n-1)\gamma} (V+\{\frac{n+(n-1)\gamma}{2n+(2n-1)\gamma}\}\gamma\bar{c}) + \frac{(n+(n-1)\gamma)\bar{c}}{2n+(2n-1)\gamma} \quad (2.37)$$

using Q from (2.35) and substituting for Δ its value $\frac{2(1+\gamma)}{\gamma}$.

Let us now compare the two equilibrium prices given by eq^n. (2.33) for Cournot game and eq^n.(2.37) for Edgeworth game.

For n = 1,

$$p^i \text{ (Cournot)} = \frac{1+\gamma}{2(1+\gamma)} (V+ \frac{\gamma}{2+\gamma} \bar{c}) + \frac{\bar{c}}{2+\gamma} = \frac{1}{2}(V+\bar{c})$$

$$p^i \text{ (Edgeworth)} = \frac{1}{2} (V+ \frac{\gamma}{2+\gamma} \bar{c}) + \frac{\bar{c}}{2+\gamma} = \frac{1}{2} (V+\bar{c})$$

For $n \to \infty$,

$$p^i \text{ (Cournot)} = \frac{1}{2+\gamma} (V+\frac{\gamma}{2}\bar{c}) + \frac{\bar{c}}{2+\gamma} = \frac{V}{2+\gamma} + \bar{c}(\frac{\gamma+1}{\gamma+2})$$

$$p^i \text{(Edgeworth)} = \frac{1}{2+\gamma}(V+\frac{1}{2}(\frac{1+\gamma}{1+\gamma})\gamma\bar{c}) + \frac{\bar{c}}{2+\gamma} = \frac{V}{2+\gamma} + \bar{c}(\frac{\gamma+1}{\gamma+2}) .$$

Except for these two extreme situations, p^i (Cournot) and p^i(Edgeworth) are different. For a graphical illustration of this difference, see Başar and Olsder (1982).

Example 2.3 (Oligopolistic Market with Advertising)

A different problem in an oligopolistic market has been studied by Balch (1971). There are n firms producing the same product and selling it at the same (predetermined) price p. In the preceding examples, we assumed that all the firms were interested in maximizing their respective profits. As has been persuasively argued by Galbraith (1973), large corporations may be primarily interested in increasing their market shares rather than their profits. We assume that the firms compete for their market shares on the basis of advertising. At the beginning of a marketing period (say, one day), each firm decides whether to maximize its profit or market share. The effect of this decision on the firm itself at the end of the period and on the decision of other firms have been investigated in the paper.

Let M^i be the market share of Fi at the beginning of the period, so that $\sum_{i=1}^{n} M^i = 1$. Decision variable of Fi is the advertising expenditure x^i for the period under consideration (a static framework is assumed throughout). Clearly, $x^i \geq 0$. Let c^i be the average production cost per unit for firm i and p the common price of the product. The total demand D for the product is assumed to be constant. Let $\Phi^i(x)$ be the expected market share of Fi at the end of the period, so that its expected profit is

$$\Pi^i(x) = (p-c^i)D\,\Phi^i(x) - x^i \qquad (2.38)$$

Each firm must operate under the budget constraint $\Pi^i(x) \geq 0$. Let

$$B^i(\tilde{x}^i) = \{x^i \geq 0 \mid \Pi^i(x^1,\ldots,x^{i-1},x^i,x^{i+1},\ldots,x^n) \geq 0\} \qquad (2.39)$$

where $\tilde{x}^i = (x^1,\ldots,x^{i-1},x^{i+1},\ldots,x^n)'$. Optimal strategy for Fi must lie in $B^i(\tilde{x}^i)$. The fact that $B^i(\tilde{x}^i)$ depends upon $x^1,\ldots,x^{i-1}$, $x^{i+1},\ldots,x^n$

necessitates a slight extension of the definition of Nash equilibrium solution. Suppose that Fi wants to maximize $\Psi^i(x)$ for $i = 1,\ldots,n$. A n-tuple $x^* = (x^{1^*},\ldots,x^{n^*})$ is a Nash equilibrium solution if $x^{i^*} \in B^i(x^{i^*})$ and

$$\Psi^i(x^{1^*},\ldots,x^{(i-1)^*},x^{i^*},\ldots,x^{n^*}) \geq \Psi^i(x^{1^*},\ldots,x^{(i-1)^*},x^i,x^{(i+1)^*},\ldots,x^{n^*}) \qquad (2.40)$$

for all $x^i \in B^i(\tilde{x}^{i^*})$ and for all $i = 1,\ldots,n$.

Each firm decides at the outset whether to maximize $\Phi^i(x)$ or $\Pi^i(x)$. Let A be the subset of firms maximizing market shares and the complementary set B consists of firms maximizing profits. We have yet to specify $\Phi^i(x)$. Let θ be the fraction of consumers who buy the product of a firm mainly on the basis of its advertisement appeal, so that $(1-\theta)$ is the fraction of consumers buying the product of a firm on the basis of previous habit. Let $\alpha^i > 0$ reflect advertising appeal of Fi. Then we take

$$\Phi^i(x) = (1-\theta)M^i + \theta \frac{\alpha^i x^i}{\sum_i \alpha^i x^i} \qquad (2.41)$$

The game problem is to determine a Nash equilibrium solution $x^* = (x^{1^*},\ldots,x^{n^*})$ when firms in A maximize $\Phi^i(x)$ and those in B maximize $\Pi^i(x)$.

<u>Case 1</u> $|A| > 1$ ($|A|$ denotes the number of elements in A)

Suppose that at least two firms advertise. This means that, we must constrain our Nash equilibrium solution to lie in

$$Q = \{x \mid x^i \geq 0, \quad \sum_{\substack{j \\ j\neq i}} \alpha^j x^j > 0 \text{ for all } i\}$$

For $x \in Q$, $\Pi^i(x)$ as a function only of the i-th component x^i is strictly concave, continuously differentiable, is positive for $x^i = 0$ and negative for large x^i. This implies that $B^i(\tilde{x}^i) = [0,\xi^i(\tilde{x}^i)]$, where $\xi^i(\tilde{x}^i)$ is the unique zero of $\Pi^i(x)$, considered as a function of x^i. Let

$\mu^i(\tilde{x}^i)$ be the unique maximum of $\Pi^i(x)$ w.r.t. x^i in $[0,\xi^i(\tilde{x}^i)]$. On the other hand, $\Phi^i(x)$ is strictly increasing w.r.t. x^i in $[0,\xi^i(\tilde{x}^i)]$ and attains its maximum at $\xi^i(\tilde{x}^i)$. It follows that $x \in Q$ is a Nash equilibrium solution for the game if and only if

$$x^i = \begin{cases} \xi^i(\tilde{x}^i) & \text{for } i \in A \\ \mu^i(\tilde{x}^i) & \text{for } i \in B \end{cases} \tag{2.42}$$

Further analysis shows that there is a unique point $x \in Q$ satisfying (2.42). Thus, if $|A| \neq 1$, the game under consideration has a unique Nash equilibrium solution.

Case 2 $|A| = 1$; say, $A = \{m\}$.

In this case, we must constrain our Nash solution to lie in $P^m = \{x | x^i \geq 0 \text{ for all } i \text{ and } x^m > 0\}$.

For $i \in B$, the previous result of having a unique x^i satisfying $x^i = \mu^i(\tilde{x}^i)$ still holds. However, $\Phi^m(x)$ is no longer strictly increasing w.r.t. x^m and among all the maximizersof $\Phi^m(x)$ w.r.t. x^m, choose that one as our solution which also maximizes $\Pi^m(x)$ w.r.t. x^m. With this kind of lexiographic ordering, we can again show that, if $|A| = 1$, the game under consideration has a unique Nash equilibrium solution.

We denote the unique Nash equilibrium solution by $x^*(A)$. We normalize α so that $\sum_i \alpha_i = 1$. Define $\sigma^i = \alpha^i(p-c^i)D\theta$ and relabel the firms so that $\sigma^1 \geq \sigma^2 \geq \ldots \geq \sigma^n > 0$. Define

$$\sigma(A) = \begin{cases} \max_i . \sigma^i & \text{if } A \neq \emptyset \\ 0 & \text{otherwise} \end{cases} \tag{2.43}$$

In terms of these parameters, one can obtain sufficient conditions under which definitive statements can be made regarding the expected loss or gain of market shares for different firms at the Nash equilibrium solution. The analysis is straightforward and the results may be summarized as follows:

Proposition (Balch, 1971)

If $A = \{m\}$ and $\sigma^m(1 + \frac{1-\theta}{\theta} M^m) > \max_{i \neq m} \sigma^i$, $\phi^i(x^*(A)) < M^i$ for all $i \in B$. If $|A| \geq 2$ or $|A| = 1$ but the preceding condition does not hold, $\phi^i(x^*(A)) < M^i$ for those $i \in B$ for which $(1-M^i)\sigma^i < \sigma(A)$. If $|A| = 0$ and $\sum_{i=1}^{n-1} \frac{\sigma^n}{\sigma^i} > n-2$, $x^*(\phi) > 0$ and $\phi^i(x^*(\phi)) < M^i$ for those i for which $(1-M^i)\sigma^i < \max_{j \neq i} \sigma^j(1-M^j)$. If $\sigma^i > \theta\sigma^1$ for all i, $\phi^i(x^*(A)) > M^i$ for those $i \in A$ for which $\sigma^i = \sigma(A)$.

Example 2.4 (Choice of Technology)

We go back to the Cournot duopoly model where the firms operating with the existing technology are confronted with the choice of adopting a new technology for producing their goods. There is uncertainty cocerning cost reduction as a result of adopting the new technology and we are faced with a game having imperfect information. There are only two pure strategies available to both the firms: 0 (continuing with old technology) and 1 (changing over to new technology). We bring in mixed strategies by defining a strategy for firm i as a number $p^i \in [0,1]$, where p^i stands for the probability that Fi will adopt the new technology.

Suppose that the present unit production cost of Fi is m^i generating profits at the rate $r^i(m^1,m^2)$. Let β_i be the rate of discount for Fi, $i = 1,2$. Firm i is uncertain about the actual unit production cost using the new technology and we take this to be a random variable c^i having a distribution function $F^i(\cdot)$ with support in $[\underline{c}^i,\bar{c}^i]$. If $\underline{c}^1 = \underline{c}^2$, $\bar{c}^1 = \bar{c}^2$ and $F^1(\cdot) = F^2(\cdot)$, we say that the uncertainty is innovation specific. Otherwise, the uncertainty is termed firm specific. Suppose that F^i is strictly increasing on $[\underline{c}^i,\bar{c}^i]$ with density function f^i and the current cost $m^i \in [\underline{c}^i,\bar{c}^i]$, $i = 1,2$. Let k^i denote the cost of adopting the new technology for Fi, $i = 1,2$.

The payoff (future profits) to Fi under the strategy pair (p^1,p^2) is denoted by $\Pi^i(p^1,p^2)$. Then

$$\Pi^i(p^1,p^2) = p^1p^2\,\Pi^i(1,1)+(1-p^1)p^2\,\Pi^i(0,1) + p^1(1-p^2)\Pi^i(1,0) + (1-p^1)(1-p^2)\Pi^i(0,0):\ i = 1,2 \quad (2.44)$$

Thus, the payoffs of the firms are specified once the payoffs corresponding to the four pure strategy pairs are known. The payoffs are exclusive of current revenues and therefore, the pure strategy payoffs for F1 are

$$\Pi^1(0,0) = \frac{\beta_1}{1-\beta_1}\, r^1(m^1,m^2) \quad (2.45a)$$

$$\Pi^1(1,0) = \frac{\beta_1}{1-\beta_1}\int_{\underline{c}^1}^{m^1} r^1(c^1,m^2)f^1(c^1)dc^1 + \frac{\beta_1}{1-\beta_1}(1-F^1(m^1))r^1(m^1,m^2)-k^1 \quad (2.45b)$$

$$\Pi^1(0,1) = \frac{\beta_1}{1-\beta_1}\int_{\underline{c}^2}^{m_2} r^1(m^1,c^2)f^2(c^2)dc^2 + \frac{\beta_1}{1-\beta_1}(1-F^2(m^2))r^1(m^1,m^2) \quad (2.45c)$$

$$\begin{aligned}\Pi^1(1,1) = {} & \frac{\beta_1}{1-\beta_1}\int_{\underline{c}^1}^{m^1}\int_{\underline{c}^2}^{m^2} r^1(c^1,c^2)f^1(c^1)f^2(c^2)dc^1dc^2 \\ & + \frac{\beta_1}{1-\beta_1}(1-F^1(m^1))\int_{\underline{c}^2}^{m^2} r^1(m^1,c^2)f^2(c^2)dc^2 \\ & + \frac{\beta_1}{1-\beta_1}(1-F^2(m^2))\int_{\underline{c}^1}^{m^1} r^1(c^1,m^2)f^1(c^1)dc^1 \\ & + \frac{\beta_1}{1-\beta_1}(1-F^1(m^1))(1-F^2(m^2))r^1(m^1,m^2)-k^1 \end{aligned} \quad (2.45d)$$

The above "technology adoption game" model has been formulated and analyzed in detail by Reinganum (1983). In the static set-up as described here, a rather involved analysis leads to the following conclusion:

<u>Proposition</u> (Reinganum, 1983).

Under the following assumptions

(A1) $\frac{\beta_1}{1-\beta_1}\int_{\underline{c}^1}^{\bar{c}^1} [r^1(c^1,\underline{c}^2)-r^1(\bar{c}^1,\underline{c}^2)]f^1(c^1)dc^1-k^1 > 0$

and

(A2) $\partial r^i/\partial m^i < 0$, $\partial r^i/\partial m^j > 0$ $(i \neq j)$ and $\partial^2 r^i/\partial m^1\partial m^2 < 0$, $i = 1,2$, a Nash equilibrium solution for the game always exists in the class of pure strategies. For some values of m, k, β and some distribution functions F, there exist two pure strategy equilibria and one mixed strategy equilibrium (p^{1^*}, p^{2^*}) given by

$$p^{1^*} = [\pi^2(0,1) - \pi^2(0,0)] / [\pi^2(0,1) - \pi^2(0,0) - \pi^2(1,1) + \pi^2(1,0)] \tag{2.46a}$$

and

$$p^{2^*} = [\pi^1(1,0) - \pi^1(0,0)] / [\pi^1(1,0) - \pi^1(0,0) - \pi^1(1,1) + \pi^1(0,1)] \tag{2.46b}$$

<u>Example 2.5</u> (Two Level Planning)

We now turn our attention to system with hierarchical structures where the natural equilibrium solution concept is that proposed by von Stackelberg. Consider a two-level planning problem consisting of one central decision maker (the State) and n decentralized units (Provinces) Each province is involved in various economic activities which we classify into n sectors. The central decision maker wants to induce appropriate sectoral activities in the provinces by providing grants for this purpose. Let u_j, $j = 1,\ldots,m$, be some sectoral activity levels desired by the central DM. Let a_j^i denote the level of demographic and socio-economic variables for sector j in province i. Then the central DM allocates grant to the amount of $f_j(a_j^i, u_j)$ to promote activity in sector j of province i. Total grant appropriated to province i is then

$$y^i = \sum_{j=1}^{m} f_j(a_j^i, u_j), \quad i = 1,\ldots,n \tag{2.47}$$

The provinces have discretion in using the grant to achieve their respective goals. Let u_j^i denote the activity level of sector j in province i, and they are the strategies for province $i, i = 1,\ldots,n$. Province i wants to maximize its own criterion (welfare)

$$W^i(\underline{u}^i, y^i), \quad \underline{u}^i = (u_1^i, \ldots, u_m^i)$$

subject to appropriate constraints. The central DM, on the other hand, wants to maximize

$W(\underline{u},\underline{u}^1,\ldots,\underline{u}^n)$, $\underline{u} = (u_1,\ldots,u_m)$

subject to the budget constraint (from eq^n. (2.47)):

$$\sum_{i=1}^{n} \sum_{j=1}^{m} f_j(a_j^i,u_j) \leq B \qquad (2.48)$$

The natural equilibrium solution for this two-level game is obvious. Thus, given $\underline{u}$ satisfying the constraint (2.48), the players at the lower level (provinces) maximize $W^i(\underline{u}^i,y^i)$. The optimal $\underline{u}^i$ will be functions of $\underline{u}$ for all $i = 1,\ldots,n$. Substituting these in $W(\underline{u},\underline{u}^1,\ldots,\underline{u}^n)$, the criterion for the central DM becomes a function of $\underline{u}$ only. Maximizing this function gives the optimal strategies $\underline{u}^*$ for the central decision maker. The optimal $\underline{u}^{i^*}$ for the provinces may now be readily obtained. The solution $(\underline{u}^*,\underline{u}^{1^*},\ldots,\underline{u}^{n^*})$ is precisely the Stackelberg equilibrium solution for this game. See Togsverd (1975) for algorithmic aspects of solving these planning problems.

Example 2.6 (Urban Planning)

We consider a somewhat more complicated two-level planning problem where the strategies of various levels are connected by means of problem constraints. This arises in urban planning and is discussed in detail in Roy and Lesse (1983).

Consider an urban community where an urban authority has to decide on allocating new lands for households and jobs in such a way that the transport cost incurred by the community members for commuting to their work places is the minimum possible (we assume one commuter per household). The commuters, on the other hand, want to determine their travel plans based on trade-offs among trip costs, trip times and relative benefits for different destinations. There is, therefore, a conflict in the objective of the authority (leader) and the members of the community (follower).

Suppose that x_i is the number of households and y_i the number of jobs in the new zone i. Let T_{ij} denote the number of trips between the zones i and j. The urban authority wants to choose x_i and y_i for all i (indirectly) so as to minimize the total travel costs

$$J^1 = \sum_{i,j} T_{ij} c_{ij} \qquad (2.49)$$

where c_{ij} denotes unit travel cost, subject to the obvious constraint

$$\sum_i x_i = \sum_j y_j = T \qquad (2.50)$$

where T stands for the total number of households (or, total number of jobs). For making the model more realistic, we must impose an additional constraint

$$x_i/d_h + y_i/d_e \leq Z_i \text{ for all } i \qquad (2.51)$$

where d_h and d_e stand for housing and employment densities, respectively and Z_i is the available area in zone i.

The number of trips T_{ij} are the decision variables for the commuters (followers). They are related to the decision variables x_i and y_i for the urban authority (leader) by means of the obvious constraints

$$\sum_j T_{ij} = x_i \quad , \quad \sum_i T_{ij} = y_j \qquad (2.52)$$

The objective for the commuters is a more complicated question that belongs to the realm of behavioral science. A commonly used objective to be maximized is the commuter entropy

$$J^2 = - \sum_{i,j} T_{ij}(\log T_{ij} + \beta g_{ij}) \qquad (2.53)$$

where β is the impedance parameter and g_{ij} is the perceived generalized travel cost between zones i and j. Thus, the commuters want to determine T_{ij}'s that maximize J^2 subject to the constraints (2.52). It is only through the constraints that the decision variables of the urban authority enter into the optimal T_{ij}'s. The solution procedure is apparent and correspond to the Stackelberg equilibrium solution concept. Given the decision x_i and y_j of the urban authority satisfying the constraints (2.50) and (2.51), the commuters determine T_{ij}'s that maximize J^2 subject to (2.53). The optimal T_{ij}'s depend upon x_i's and y_j's. Substituting these optimal T_{ij}'s in J^1, we can express J^1 as a function of x_i's and y_j's. The central authority then minimizes J^1 to determine the optimal x_i's and y_j's which, in turn, determine the optimal T_{ij}'s.

To maximize J^2 subject to (2.52), we form the Lagrangian

$$L = -\sum_{i,j} T_{ij}(\log T_{ij} + \beta g_{ij}) + \sum_i \mu_i (x_i - \sum_j T_{ij}) + \sum_j \eta_j (y_j - \sum_i T_{ij}) \tag{2.54}$$

The optimal T_{ij}'s satisfy $\partial L / \partial T_{ij} = 0$. This yields

$$\log T_{ij} + 1 + \beta g_{ij} + \mu_i + \eta_j = 0.$$

Writing $X_i = e^{-\mu_i}$, $Y_j = e^{-\eta_j}$ and $r_{ij} = e^{-1-\beta g_{ij}}$, we get

$$T_{ij} = X_i Y_j \, r_{ij} \tag{2.55}$$

Using $\sum_j T_{ij} = x_i$ and $\sum_i T_{ij} = y_j$, we can write

$$T_{ij} = x_i y_j \, A_i B_j r_{ij} \tag{2.56}$$

where

$$A_i = (\sum_j B_j y_j r_{ij})^{-1} \text{ and } B_j = (\sum_i A_i x_i r_{ij})^{-1} \tag{2.57}$$

Subctituting this in (2.49), we get

$$J^1 = \sum_{i,j} x_i y_j [A_i B_j r_{ij} c_{ij}] \tag{2.58}$$

The urban authority determines optimal x_i's and y_j's by minimizing J^1 given by (2.58) subject to the constraints (2.50) and (2.51). The optimal T_{ij}'s are obtained by substituting these in (2.56).

The expression for J^1 in (2.58) is extremely involved and computing the gradient for minimizing J^1 is numerically very complicated. An alternative approach is, instead, to work with T_{ij} given by (2.55). This expression for T_{ij} is directly substituted into (2.49) to express

J^1 in terms of X_i's and Y_j's. The constraints (2.50) and (2.51) are also similarly transformed into constraints in terms of X_i's and Y_j's. The objective is now quadratic (or bilinear) and the constraints are also quadratic (or bilinear). It is an equivalent optimization problem and can be solved more easily. Details may be found in Roy and Lesse (1983).

CHAPTER 3

MATHEMATICAL FORMULATION OF DIFFERENTIAL GAMES

1. INTRODUCTION

In this chapter we formulate differential game problems and explain in detail both the Nash and the Stackelberg "equilibrium solution" concepts. Such games may be defined in the discrete-time as also in the continuous-time and both the cases will be carefully analyzed. We shall also mention mixed strategies and stochastic differential games without getting into details. Differential games, of course, involve a multiple number of players or decision makers. To solve the general problem, one has to often consider a number of degenerate problems involving only one player in which case one is confronted with the classical optimal control problem. Necessary conditions for solving this problem, the so-called Pontryagin Maximum Principle, is explained from an economic viewpoint. Finally, the use of the maximum principle is illustrated by means of an example arising in the regional investment allocation problem.

2. CONTINUOUS-TIME DIFFERENTIAL GAMES

We start with the continuous-time formulation. We take an informal point of view and present the essential aspects of a differential game without pursuing an axiomatic approach. We consider a dynamical system described by a set of ordinary differential equations

$$\frac{dx(t)}{dt} = f(t,x(t),\ u^1(t),\ldots,u^N(t)),\ t \in [0,T] \tag{3.1}$$

$$x(0) = x_0 \text{ , given}$$

where $x(t) \in \mathbb{R}^n$ and $f = \begin{bmatrix} f_1 \\ \\ f_n \end{bmatrix}$.

The vector $x(t)$ is called the state of the system at time t and x_0 is called the initial state. The evolution of the state is influenced or controlled by N agents via the controls $u^1,\dots,u^N$. We shall call them players or decision makers (D.M. for short). The vectors $u^i(t)$, for each t, belong to $\mathbb{R}^{m_i}$, $i = 1,\dots,N$. We assume throughout that

(a) $(t,x,u^1,\dots,u^N) \to f(t,x,u^1,\dots,u^N)$ is continuous from

$$[0,T] \times \mathbb{R}^n \times \mathbb{R}^{m_1} \times \dots \times \mathbb{R}^{m_N} \to \mathbb{R}^n$$

(b) $(t,x,u^1,\dots,u^N) \to \frac{\partial f_i}{\partial x_j}(t,x,u^1,\dots,u^N)$ is continuous

$$\text{from } [0,T] \times \mathbb{R}^n \times \mathbb{R}^{m_1} \times \dots \times \mathbb{R}^{m_N} \to \mathbb{R}.$$

In practice, one has to restrict the controls further. Thus, let U^i_{ad} be a subset of $\mathbb{R}^{m_i}$. An <u>admissible control</u> $u^i(\cdot)$, $i = 1,\dots,N$, is a piecewise continuous mapping from $[0,T] \to U^i_{ad}$, such that eq^n. (3.1) has a unique solution. We do not assume any constraint on the state of the system. Introducing state constraint makes the optimization problem one is interested in a very difficult one.

The optimization problem means maximizing or minimizing some criterion. If now there is one single criterion that all the players want to optimize together, one is faced with a standard optimal control problem. Differential games differ from deterministic optimal controls in essentially two respects. Firstly, each player wants to optimize his own criterion and the different criteria for different players may well be in conflict with one another. Secondly, controls at each time may be specified via the information available to the players about the state of the system until that time. These informations may be different for different players and, unlike in the optimal control problems, different optimal solutions will be obtained with different information structures in differential games. Let us now specify them in mathematical terms.

We assume that player i wants to maximize his own criterion or pay of ("welfare" in many economic problems)

$$W^i(u^1,\dots,u^N) = \int_0^T \phi^i(t,x(t),u^1(t),\dots,u^N(t))dt + g^i(x(T)) \qquad (3.2)$$

for differentiable functions $\phi^i(t,\cdot,u^1,\ldots,u^N)$ and $g^i(\cdot)$, $i = 1,\ldots,N$. If the criterion has to be minimized, one can work with its negative value which will then have to be maximized. Furthermore, controls $u^i(t)$, for each t, are specified by means of information structures. Thus, suppose that player i has available, at time t, information about the state x(s), $0 \le s \le \tau^i(t)$; $0 \le \tau^i(t) \le t$ (for causality), or its discrete counterpart, $x(s_1^i(t)),\ldots,x(s_p^i(t))$; $s_1^i(t) \le \ldots \le s_p^i(t) \le t$ (also for causality). Depending on the type of information available, we can define <u>strategy space</u> Γ^i of player i of all suitable mappings as follows:

(a) $\gamma^i(t,x(\cdot))$, where for fixed t, γ^i depends on x(s) for s in $[0,\tau^i(t)]$. This can, of course, be formalized using measure theory. We also require $u^i(t) = \gamma^i(t,x(\cdot))$ to belong to U^i_{ad} for $i = 1,\ldots,N$.

or

(b) $\gamma^i(t,x(s_1^i(t)),\ldots,x(s_p^i(t)))$ as a mapping from $[0,T] \times \mathbb{R}^{pn} \to U^i_{ad}$, $i = 1,\ldots,N$.

One may consider more complex information structures. For example, in case (b) above, p may be allowed to depend on t and/or, we may have a combination of cases (a) and (b) as our available information. These situations have hardly any practical relevance. The case of partial or incomplete observation of state is, of course, of considerable practical interest. We discuss this explicitly later for discrete-time systems.

The mapping γ^i introduced above in cases (a) and (b) are constrained by the fact that if we replace $u^i(t)$ in eq^n. (3.1) by $\gamma^i(t,x(\cdot))$ or $\gamma^i(t,x(s_1^i(t)),\ldots,x(s_p^i(t)))$, the resulting differential equation should have a unique solution x. This solution, put back into the form for γ^i, yields the desired control actions $u^i(t)$, $i = 1,\ldots,N$. Conditions have to be imposed on f and the class Γ^i to ensure unique

solution of eq^n. (3.1).
Lipschitz condition is obviously the only generally available sufficient condition to ensure uniqueness of solution of eq^n. (3.1). But Lipschitz condition may be too demanding in a N-person differential game. In a 2-person game, if one player would restrict his strategy to be Lipschitzian, the other player may be able to exploit it. This is clearly exhibited in "pursuit-evasion" games. Unfortunately, however, non-Lipschitzian strategy cannot be easily handled mathematically.

Let us now define equilibrium concepts for the multi-criteria optimization problem formulated above. For this, the role of strategies are brought more explicitly into the picture. For a fixed N-tuple of strategies $\gamma^i \in \Gamma^i$ of the type (a) or (b) above, $i = 1,\ldots,N$, suppose that eq^n. (3.1) has a unique solution. This solution, in turn, determines uniquely controls $u^i(t)$, $0 \leq t \leq T$. Substituting the trajectory and the N control functions into our criteria W^i, we get a fixed number for each i for $i = 1,\ldots,N$. Thus, for each fixed i, we get a mapping

$$J^i : \Gamma^1 X \ldots X \Gamma^N \to \mathbb{R}.$$

We, therefore, write the optimizing criteria for player i as $J^i(\gamma^1,\ldots,\gamma^N)$, $i = 1,\ldots,N$. It is in this form that we define Nash and Stackelberg equilibrium solutions for the differential game problem formulated here.

Definition 3.1

A N-tuple of strategies $\{\gamma^{1^*},\ldots,\gamma^{N^*}\}$ with $\gamma^{i^*} \in \Gamma^i$, $i = 1,\ldots,N$, is called a Nash equilibrium solution for the differential game under consideration if

$$\begin{aligned} J^1(\gamma^{1^*},\gamma^{2^*},\ldots,\gamma^{N^*}) &\geq J^1(\gamma^1,\gamma^{2^*},\ldots,\gamma^{N^*}) \\ &\vdots \\ J^N(\gamma^{1^*},\gamma^{2^*},\ldots,\gamma^{N^*}) &\geq J^N(\gamma^{1^*},\gamma^{2^*},\ldots,\gamma^N) \end{aligned} \tag{3.3}$$

Let $J^{i^*} \triangleq J^i(\gamma^{1^*}, \gamma^{2^*}, \ldots, \gamma^{N^*})$. Then $\{J^{1^*}, \ldots, J^{N^*}\}$ is called a <u>Nash equilibrium outcome</u> of the game.

Intuitively, Nash equilibrium concept means that if one player tries to alter his strategy unilaterally, he cannot improve his own optimization criterion or welfare. There is clear symmetry in the role of the players in this sort of situation. The solution or equilibrium concept is, therefore, reasonable in the situation where no player or DM has any special role in the decision making process. The game is also played noncooperatively with each player only interested in his own optimization criterion. It is quite possible for all players to improve their criteria by playing cooperatively among them.

There is another solution or equilibrium concept in differential games which become relevant in the situation where one or several players has (have) dominant role(s) vis-a-vis the rest of the players. Following the original work of H. von Stackelberg (1934), player(s) who holds the dominant position in such a game is (are) called leader(s) and the rest are called followers. Let us explain this Stackelberg equilibrium concept intuitively for the case of two-player game with player P1 as the leader and player P2 as the follower. P1 can impose his decision on P2. Suppose that P1 chooses a strategy γ^1. P2 has to maximize his own optimization criterion as a reaction to the announced strategy γ^1 of P1. Assume that for each γ^1, there is a <u>unique</u> optimal strategy γ^2 for P2. Thus, there is a mapping $T : \Gamma^1 \to \Gamma^2$ such that $T(\gamma^1) = \gamma^2$ is the unique optimal strategy of P2 in response to the announced strategy γ^1 of P1. The optimal strategy γ^{1^*} in this context is now obvious. It is a strategy such that

$$J^1(\gamma^{1^*}, T(\gamma^{1^*})) \geq J^1(\gamma^1, T(\gamma^1)) \forall \gamma^1 \in \Gamma^1 \tag{3.4}$$

and clearly $\gamma^{2^*} = T(\gamma^{1^*})$ is the optimal strategy for P2. In general, this uniqueness assumption of the optimal strategy for P2 in response to any strategy announced by P1 may not hold. Furthermore, for a general N-person game, numerous combinations for leaders and followers are possible. A comprehensive definition, therefore, becomes difficult to write down. Below we give the definition of Stackelberg equilibrium solution first for two players and then for three players with all possible combinations of leaders and followers.

Definition 3.2 (N = 2)

The optimal reaction set $R^2(\gamma^1)$ of P2 to the strategy $\gamma^1 \in \Gamma^1$ of P1 is defined by

$$R^2(\gamma^1) = \{\zeta \in \Gamma^2 | J^2(\gamma^1,\zeta) \geq J^2(\gamma^1,\gamma^2) \quad \forall \gamma^2 \in \Gamma^2\}.$$

The uniqueness assumption above refers to the situation when $R^2(\gamma^1)$ is a singleton. In a two-person differential game with P1 as the leader, a strategy $\gamma^{1^*} \in \Gamma^1$ is called a Stackelberg equilibrium solution for the leader if

$$\min_{\gamma^2 \in R^2(\gamma^{1^*})} J^1(\gamma^{1^*},\gamma^2) \geq \min_{\gamma^2 \in R^2(\gamma^1)} J^1(\gamma^1,\gamma^2) \quad \forall \gamma^1 \in \Gamma^1 \tag{3.5}$$

This is clearly a generalization of (3.4). If J^{1^*} denotes the l.h.s. of the above inequality (3.5), we call J^{1^*} the Stackelberg payoff of the leader. If $\gamma^{1^*} \in \Gamma^1$ is a Stackelberg solution for the leader P1, any element $\gamma^{2^*} \in R^2(\gamma^{1^*})$ is an optimal strategy for the follower P2 that is in equilibrium with γ^{1^*}. The pair $\{\gamma^{1^*},\gamma^{2^*}\}$ is a Stackelberg equilibrium solution with P1 as the leader and $\{J^1(\gamma^{1^*},\gamma^{2^*}), J^2(\gamma^{1^*},\gamma^{2^*})\}$ is the corresponding Stackelberg equilibrium outcome of the game.

Definition 3.3 (N = 3)

We consider three cases separately.

(a) We consider a 3-person differential game with one leader P1 and two followers P2 and P3. This is a typical set-up for a hierarchical decision making problem. The optimal reaction set $R^F(\gamma^1)$ of the followers to the strategy $\gamma^1 \in \Gamma^1$ of the leader is defined by

$$R^F(\gamma^1) = \{(\zeta^2,\zeta^3) \in \Gamma^2 X \Gamma^3 | J^2(\gamma^1,\zeta^2,\zeta^3) \geq J^2(\gamma^1,\gamma^2,\zeta^3),$$
$$J^3(\gamma^1,\zeta^2,\zeta^3) \geq J^3(\gamma^1,\zeta^2,\gamma^3), \quad \forall\gamma^2 \in \Gamma^2, \gamma^3 \in \Gamma^3\}.$$

In this 3-person differential game, $\gamma^{1^*} \in \Gamma^1$ is a <u>hierarchical equilibrium solution</u> for the leader if

$$\min_{(\gamma^2,\gamma^3) \in R^F(\gamma^{1^*})} J^1(\gamma^{1^*},\gamma^2,\gamma^3) \geq \min_{(\gamma^2,\gamma^3) \in R^F(\gamma^1)} J^1(\gamma^1,\gamma^2,\gamma^3) \qquad \forall\gamma^1 \in \Gamma^1 \tag{3.6}$$

Any $(\gamma^{2^*},\gamma^{3^*}) \in R^F(\gamma^{1^*})$ is a <u>corresponding optimal strategy pair</u> for the followers.

(b) We consider a 3-person differential game with two leaders P1 and P2 and one follower P3. The <u>optimal reaction set</u> $R^3(\gamma^1,\gamma^2)$ of the follower to the strategy $(\gamma^1,\gamma^2) \in \Gamma^1 X \Gamma^2$ of the leaders is defined by

$$R^3(\gamma^1,\gamma^2) = \{\zeta \in \Gamma^3 | J^3(\gamma^1,\gamma^2,\zeta) \geq J^3(\gamma^1,\gamma^2,\gamma^3) \quad \forall\gamma^3 \in \Gamma^3\}.$$

In this 3-person differential game, $(\gamma^{1^*},\gamma^{2^*}) \in \Gamma^1 X \Gamma^2$ is a <u>hierarchical equilibrium solution</u> for the leaders if

$$\begin{aligned}
&\min_{\gamma^3 \in R^3(\gamma^{1^*},\gamma^{2^*})} J^1(\gamma^{1^*},\gamma^{2^*},\gamma^3) \geq \min_{\gamma^3 \in R^3(\gamma^1,\gamma^{2^*})} J^1(\gamma^1,\gamma^{2^*},\gamma^3) \qquad \forall\gamma^1 \in \Gamma^1 \\
&\min_{\gamma^3 \in R^3(\gamma^{1^*},\gamma^{2^*})} J^2(\gamma^{1^*},\gamma^{2^*},\gamma^3) \geq \min_{\gamma^3 \in R^3(\gamma^{1^*},\gamma^2)} J^2(\gamma^{1^*},\gamma^2,\gamma^3) \qquad \forall\gamma^2 \in \Gamma^2
\end{aligned} \tag{3.7}$$

Any strategy $\gamma^{3^*} \in R^3(\gamma^{1^*},\gamma^{2^*})$ is a <u>corresponding optimal strategy</u> for the follower. It is clear that in cases (a) and (b) we have defined equilibrium solution concept such that Stackelberg concept comes in between hierarchies while Nash concept is used for equilibrium solution between players in the same level (leaders or followers).

(c) Consider a 3-person differential game with hierarchy between all of them (three levels of hierarchy) with P1 at the highest level, P2 at the middle level and P3 at the lowest level. In this case, we need to define two optimal reaction sets $R^3(\gamma^1,\gamma^2)$ and $R^2(\gamma^1)$ as follows:

$$R^3(\gamma^1,\gamma^2) = \{\zeta \in \Gamma^3 | J^3(\gamma^1,\gamma^2,\zeta) \geq J^3(\gamma^1,\gamma^2,\gamma^3) \quad \forall \gamma^3 \in \Gamma^3\}$$

$$R^2(\gamma^1) = \{\zeta \in \Gamma \,|\, \min_{\gamma^3 \in R^3(\gamma^1,\zeta)} J^2(\gamma^1,\zeta,\gamma^3) \geq \min_{\gamma^3 \in R^3(\gamma^1,\gamma^2)} J^2(\gamma^1,\gamma^2,\gamma^3) \quad \forall \gamma^2 \in \Gamma^2\}$$

In this 3-person differential game, $\gamma^{1^*} \in \Gamma^1$ is a <u>hierarchical equilibrium solution</u> for P1 if

$$\min_{\gamma^2 \in R^2(\gamma^{1^*})} \; \min_{\gamma^3 \in R^3(\gamma^{1^*},\gamma^2)} J^1(\gamma^{1^*},\gamma^2,\gamma^3) \geq \min_{\gamma^2 \in R^2(\gamma^1)} \; \min_{\gamma^3 \in R^3(\gamma^1,\gamma^2)} J^1(\gamma^1,\gamma^2,\gamma^3) \quad \forall \gamma^1 \in \Gamma^1 \tag{3.8}$$

Any element $\gamma^{2^*} \in R^2(\gamma^{1^*})$ is a <u>corresponding optimal strategy</u> for P2 and any $\gamma^{3^*} \in R^3(\gamma^{1^*},\gamma^{2^*})$ is an <u>optimal strategy</u> for P3 corresponding to $(\gamma^{1^*},\gamma^{2^*})$.

<u>Remark</u>

The above definition is extremely messy and considerable simplification is achieved when the various reaction sets appearing in the definition turn out to be singletons.

Let us now turn our attention to the information structure. We previously introduced a general notion of information available to the players. Further specification of information pattern leads to different kinds of differential games. We discuss them in the context of the deterministic set-up we have considered so far.

Definition 3.4

Let us recall two types of general information structures (a) and (b) discussed previously. In a N-person continuous-time differential game of fixed duration $[0,T]$, we say that Pi's information structure is

(i) open-loop (OL) pattern if strategy of type (a) $\gamma(t,x(\cdot))$ or, of type (b) $\gamma(t,x(s_1(t)),\ldots,x(s_p(t)))$ is only a function of the first (time) argument t, for all $t \in [0,T]$.

(ii) closed-loop perfect state (CLPS) pattern if we have strategy of type (a) with $\tau(t) = t$, for all $t \in [0,T]$.

(iii) ε - delayed closed-loop perfect state pattern if $\tau(t) = t - \varepsilon$, $\varepsilon > o$ fixed, for all $t \in [\varepsilon,T]$.

(iv) feedback perfect state (FB) pattern if we have strategy of type (b) with $p = 1$ and $s_1(t) = t$, for all $t \in [0,T]$.

It is possible to define other information patterns, including those with imperfect state when one has partial and/or incomplete observation of the state available. Interesting problems arise when the players optimize a single criterion but under different information patterns. They are referred to as Team problems. The subject originated with the classic work of Marschak and Radner (1972).

In general, different players may select their strategies under different information patterns. When all the players use OL-information pattern, the distinction between strategies and controls become superfluous. One can then determine the equilibrium solution with minimum complication. Optimization with respect to one control function while fixing the others form an essential step in determining solutions in such situations. This is essentially a standard optimal control problem and results from optimal control theory may be directly used to solve it.

3. DISCRETE-TIME DIFFERENTIAL (DIFFERENCE) GAMES

We now turn to the discrete-time version (more appropriately called difference games). We now have a state x_k, $k = 0,1,\ldots,K$, taking values in $\mathbb{R}^n$ and evolving according as

$$x_{k+1} = f_k(x_k, u_k^1, \ldots, u_k^N) \tag{3.9}$$

x_0 given

Here $u_k^1,\ldots,u_k^N$, $k = 0,1,\ldots,K-1$, denote the controls of the N players, with $u_k^i \in \mathbb{R}^{m_i}$ $\forall k$, $1 \le i \le N$. We have the same smoothness assumptions about f_k's as in the continuous-time case. The controls are further restricted by the constraint that $u_k^i \in U_{ad}^i \subset \mathbb{R}^{m_i}$ $\forall k$, $1 \le i \le N$. We assume now that player i wants to maximize his own welfare which we take to have the simplified "stage-additive" form:

$$W^i(u^1,\ldots,u^N) = \sum_{k=0}^{K-1} \phi_k^i(x_{k+1}, u_k^1, \ldots, u_k^N, x_k) \tag{3.10}$$

where $u^j = (u_0^{j\,'},\ldots,u_{K-1}^{j\,'})'$, for differentiable functions ϕ_k^i's. Furthermore, controls u_k^i, for each k, are specified by means of information structures. The information available to player i at time step k is a subset of $\{x_1,\ldots,x_k\}$. We denote this subset by η_k^i which characterizes information pattern of player i at time k. It is possible to consider still larger class of information by actually including the controls used by the players until the present time in the information structure. Strategy for player i now consists of all possible mappings γ_k^i from η_k^i into U_{ad}^i. Thus, $\gamma^i = \{\gamma_0^i,\ldots,\gamma_{K-1}^i\}$ represents the strategy for player i. For any permissible strategy γ^i, $i = 0,\ldots,K-1$, one can solve eq^n. (3.9) to obtain x_k, $k = 1,\ldots,K$ and use them to determine u_k^i, $k = 0,\ldots,K-1$. We note that u_0^i is determined from initial information only. We may, therefore, convert the criterion W^i into one of the form $J^i(\gamma^i,\ldots,\gamma^N)$ mapping the product strategy space $\Gamma^1 X \ldots X \Gamma^N$ into $\mathbb{R}$. In this form, we have identical definitions for Nash and Stackelberg equilibrium solutions as in the context of continuous time games. We now specify information structures further to define different kinds of games.

Definition 3.5

In a N-person discrete-time differential game of fixed duration K, we say that player Pi's information structure is

(i) open-loop (OL) pattern if $\eta_k^i = \phi$ for all k

(ii) closed-loop perfect state (CLPS) pattern if $\eta_k^i = \{x_1,\ldots,x_k\}$, $k = 1,\ldots,K-1$.

(iii) feedback perfect state (FB) pattern if $\eta_k^i = \{x_k\}$, $k = 1,\ldots,K-1$.

(iv) one-step delay CLPS (DCLPS) pattern if $\eta_k^i = \{x_1,\ldots,x_{k-1}\}$, $k = 2,3,\ldots,K-1$. In this case, η_0^i and η_1^i are both determined only on the basis of initial information.

Many variations and extensions to our discussions on differential games are possible. In defining strategies, we have so far taken γ^i having fixed function form. One may enlarge the class of strategies by allowing γ^i's to be determined by chance mechanism. These mixed strategies have been extensively studied in finite games. We do not consider them here. A more obvious extension is to allow for separate observation equations (e.g., partial observation of the state) and each player may have different observation available to him. Thus, if y_k^i be the observation of state x_k available to player i, one may consider a very useful information structure, the so-called "one-step delay observation sharing pattern", where

$\eta_k^i = \{y_1,\ldots,y_{k-1},y_k^i\}$, $k = 1,\ldots,K-1$, $y_j = \{y_j^1,\ldots,y_j^N\}$.

The assumption of fixed finite duration of the game is also restrictive. One may consider games over an infinite time-horizon and also games which terminate at a time when the state enters a target set which is specified a-priori.

We now briefly outline the formulation of stochastic difference games. In this case, the state $\{x_k,\ k = 1,\dots,K\}$ becomes a stochastic process whose evolution is characterized by

$$x_{k+1} = f_k(x_k, u_k^1, \dots, u_k^N, n_k) \tag{3.11}$$

x_0 given and fixed.

The additional argument n_k, $k = 0,\dots,K-1$, form a sequence of random vectors which are usually taken to be independent (white). The above equation expresses that if the system at time k is in a given state x_k and controls $u_k^1,\dots,u_k^N$ are applied, the next state x_{k+1} will be a random vector, whose distribution depends on k, x_k and $u_k^1,\dots,u_k^N$. The welfare to be maximized by player Pi may be taken, as in the deterministic case, to be

$$\tilde{W}^i(u^1,\dots,u^N) = \sum_{k=0}^{K-1} \phi_k^i(x_{k+1}, u_k^1, \dots, u_k^N, x_k),\ i = 1,\dots,N$$

and our objective now is to find a method for determining how the controls $u_k^1,\dots,u_k^N$ should be chosen from one time point to the next depending upon the information available at each instant, such that a suitable equilibrium solution for the game is obtained. But because of the random nature of the model, it is impossible to predict what the precise outcome of the total cost will be before the control realizations actually occur. Therefore, the welfare function to be maximized by Pi is not $\tilde{W}^i$ but its expectation:

$$W^i(u^1,\dots,u^N) = E\{\sum_{k=0}^{K-1} \phi_k^i(x_{k+1}, u_k^1, \dots, u_k^N, x_k)\} \tag{3.12}$$

To define equilibrium solutions, we have to again convert welfare functions into the form where the arguments are elements from strategy spaces as in the deterministic case. It is possible to formulate similar games for continuous-time problems as well. The stochastic sequence $\{n_k,\ k = 0,1,\dots,K-1\}$ has then to be replaced by an appropriate stochastic process and theory of stochastic differential equations has to be used to define the state dynamics appropriately. This introduces additional complication and is omitted here.

As is obvious from the definitions, results from optimal control theory form an essential intermediate step in obtaining Nash and Stackelberg equilibrium solutions. In the next section, we briefly recapitulate results from optimal control theory we shall need in subsequent discussions and also give an economic interpretation of the Pontryagin maximum principle.

4. RESULTS FROM OPTIMAL CONTROL THEORY

Suppose that the state $x(t) \in \mathbb{R}^n$ of a dynamical system evolve according to a vector differential equation

$$\dot{x}(t) = f(t,x(t),u(t)) \ , \ 0 \le t \le T \qquad (3.13)$$
$$x(0) = x_0, \text{ given}$$

The control u(t), taking value in $\mathbb{R}^m$ for each t, is a piecewise continuous vector-valued function of time. In practice, u(t) is often constrained to lie, for each t, in a set $U_{ad} \subset \mathbb{R}^m$ which may, in general, be time varying. The problem in optimal control theory is to choose u(t), $0 \le t \le T$, satisfying the constraints mentioned above, so as to maximize

$$J(u) = \int_0^T \phi(t,x(t),u(t))\, dt + g(x(T)). \qquad (3.14)$$

We assume that $f(t,\cdot,u)$, $\phi(t,\cdot,u)$ and $g(\cdot)$ are continuously differentiable on $\mathbb{R}^n$, $\forall t \in [0,T]$. The pair $\{x(t),u(t);\ 0 \le t \le T\}$ satisfying eq^n. (3.13) and the control constraints is called a program in the economic literature and the period T is often called the planning horizon. Suppose that $\{u^*(t);\ 0 \le t \le T\}$ be a control function that maximizes J(u) subject to the constraints. Then it must satisfy a set of necessary conditions known as the Pontryagin maximum principle (see Fleming and Rishel (1975) for a proof of the maximum principle):

Theorem 3.1

Suppose that $\{x^*(t), u^*(t);\ 0 \le t \le T\}$ is an optimum program. Then there exists $n\times 1$-valued "multiplier" function $p(t)$, $0 \le t \le T$, called the costate vector such that if we define a Hamiltonian function

$$H(t,x,u,p) = \phi(t,x,u) + p'f(t,x,u) \tag{3.15}$$

the following necessary conditions hold:

$$\begin{aligned} \dot{x}_i^*(t) &= \frac{\partial H}{\partial p_i} = f_i(t,x^*(t),u^*(t)) \\ x_i(0) &= x_{0i}\ ,\ i = 1,\dots,n \end{aligned} \tag{3.16}$$

$$\dot{p}_i(t) = -\frac{\partial H}{\partial x_i}(t,x^*,u^*,p),\ i = 1,\dots,n \tag{3.17}$$

$$p_i(T) = \frac{\partial g}{\partial x_i}(x^*(T)),\ i = 1,\dots,n \tag{3.18}$$

$$H(t,x^*(t),u^*(t),p(t)) = \max_{u \in U_{ad}} H(t,x^*(t),u,p(t)) \text{ a.e.t.} \tag{3.19}$$

Remark

Just as we imposed constraints on the control function, we could impose constraints on the state as well. One could, in general, have equality or inequality constraints involving some functionals of the state and/or the control functions. These are considerably more difficult problems. Necessary conditions for optimality of one such type of constraints will be considered towards the end of this section.

Before proceeding further, let us give an economic interpretation of the maximum principle. For a masterly and detailed exposition, see Dorfman (1969). We illustrate the main economic ideas behind the maximum principle using the simplest planning problem confronting a central planning board (CPB). Consider an economy with a constant labour force and capital $K(t)$ at time t producing an output $Y(t)$ at that time, related by means of a production function F (technology) by

$$Y(t) = F(K(t)); \quad F(0) = 0 \tag{3.20}$$

The production function has "decreasing return to scale" property which means that

$$\frac{dF(K)}{dK} > 0 \text{ and } \frac{d^2F(K)}{dK^2} < 0 \quad \text{for } K \geq 0.$$

Let $C(t)$ be the consumption at time t and μ denote the rate of capital depreciation. Then the capital evolution is given by

$$\dot{K}(t) = F(K(t)) - C(t) - \mu K(t) \tag{3.21}$$

$$K(0) = K_0, \text{ initial capital}$$

This is the state evolution equation in system terminology and is simply the economic identity "savings = investment", in other contexts known also as "supply = demand principle" (market clearing). The consumption stream is related to welfare which the CPB wants to maximize. Let $u(C(t))$ denote the welfare per unit time at time t where the welfare function u has "decreasing marginal utility" property which means that

$$\frac{du(C)}{dC} > 0 \text{ and } \frac{d^2u(C)}{dC^2} < 0.$$

There is also utility attached to the final capital stock $K(T)$ at the end of the planning period $[0,T]$ which we denote by $g(K(T))$. Then the problem confronting the CPB is to choose a consumption path $C(t)$, $0 \leq t \leq T$, so as to maximize the total welfare:

$$W(C) = \int_0^T u(C(t))dt + g(K(T)) \tag{3.22}$$

The Hamiltonian at time t is defined by

$$H(K,C,p) = u(C) + p[F(K) - C - \mu K] \tag{3.23}$$

where $p(t)$, $0 \leq t \leq T$, denotes the costate variable. From the maximum principle, we know that the optimal C maximizes the

Hamiltonian. From our assumption on $u(\cdot)$, we see that

$$\frac{\partial H}{\partial C} = u_C(C) - p = 0 \tag{3.24}$$

yields a maximum of H. Further, p(t) satisfies

$$\dot{p} = - \frac{\partial H}{\partial K} = - pF_K(K) + \mu p \tag{3.25}$$

$$p(T) = \frac{dg(K(T))}{dK} \tag{3.26}$$

Let us now discuss an economic interpretation of the costate variable. Suppose that at time s, after applying the optimal control $C^*(\cdot)$ between 0 and s, which gives us the capital $K^*(s)$, we impose a slight disturbance to the capital giving us a new capital $K^*(s) + \Delta K^*(s)$ at that time. Obviously, it is no longer optimal to use $C^*(t)$ for $t > s$. Let $C^*(t) + \Delta C^*(t)$ be the new optimal consumption pattern after time s. The variation in the optimal welfare is then given by

$$\Delta W^* = \int_s^T (u(C^*(t) + \Delta C^*(t)) - u(C^*(t))\,dt + g(K^*(T) + \Delta K^*(T)) - g(K^*(T))$$

and using the optimality conditions, one can show that, in the limit (see Arrow and Kurz (1970) for details)

$$p(s) = \frac{\partial}{\partial K}\left[\int_s^T u(C^*(t))dt + g(K^*(T))\right] \tag{3.27}$$

where the partial derivative is evaluated at $K^*(s)$. The term on the right hand side of eq^n. (3.27) may be interpreted as the anticipated increase in welfare value from s to T with an addition of a unit of capital at time s. With W indicating value and K capital, p obviously has the unit of price of that capital and p(s) is the value assigned to a unit of invested capital at time s because of the anticipated welfare increase over [s,T]. Therefore, p(s) is called the <u>shadow price</u> of capital K at time s.

We have already interpreted state equation as a balance relation. Let us now interprete the costate eq^n. (3.25) under the simplifying assumption that $\mu = 0$. Then eq^n. (3.25) gives us

$$\int_s^T \dot{p}(t)dt = - \int_s^T p(t)F_K(K(t))dt$$

that is,

$$p(s) = \int_s^T p(t)F_K(K(t))dt + p(T) \qquad (3.28)$$

Eq^n. (3.28) states that we should invest capital at time s to the extent that the cumulative value of the marginal product of capital from s to T and the scrap (residual) price of capital at T together should exactly equal the shadow price of capital at time s. We can call this the "investment principle". Let us now turn to the Hamiltonian H(K,C,p). It consists of two terms. The first one, viz. u(C), stands for welfare value of current consumption and the second one, viz. $p[F(K) - C - \mu K] = p\dot{K}$, can be interpreted as the welfare value (with p indicating shadow price of capital) anticipated from the use of the capital currently added. Thus, when we maximize the Hamiltonian, we really aim at determining a consumption plan at time t given K(t) and p(t) which maximizes current welfare plus the anticipated welfare from future consumption via current investment. We can call this the "consumption principle". Thus "savings = investment principle", "investment principle" and "consumption principle" really signify eq^n. (3.16), eq^{n}s. (3.17)-(3.18) and eq^n. (3.19) of Theorem 3.1 respectively in the context of economic planning. The preceding discussion follows closely that in Isard et al (1979).

So far, we have considered the problem of optimizing J(u) with $u \in U_{OL}$, where U_{OL} is the class of piecewise continuous m×1-valued functions in [0,T] with values at each time t being constrained to lie in a set U_{ad}. Thus, our control is open-loop, is selected once and for all at the beginning of the optimization process and is independent of the actual evoluation of the system state. Similar to our discussion on information structure for differential games, we can define closed-loop and feedback control strategies as well.

Thus, let Γ be the class of functions

$\gamma : [0,T] \times \mathbb{R}^n \to U_{ad} \subset \mathbb{R}^m$ such that

$\dot{x} = f(t,x,\gamma(t,x)),\ x(0) = x_0$

has a unique solution in [0,T]. $u \in U_{FB}$, or, u is a feedback control if it has the form

$u(t) = \gamma(t,x(t))$

for some $\gamma \in \Gamma$. Mappings γ which are independent of the second argument clearly represent open-loop controls; that is, $U_{OL} \subset U_{FB}$ and so,

$$\max_{u \in U_{FB}} J(u) \geq \max_{u \in U_{OL}} J(u) \tag{3.29}$$

Suppose that γ^* maximizes J(u) for $u \in U_{FB}$. Then

$\dot{x} = f(t,x,\gamma^*(t,x))\ ,\ x(0) = x_0$

uniquely determines $x^*(t)$. If we define the open-loop control

$u^*(t) = \gamma^*(t,x^*(t)),$

then $x^*(t)$ can be recovered by solving

$\dot{x} = f(t,x,u^*),\ x(0) = x_0$

so that we can find one $u \in U_{OL}$ which will achieve the same maximum for J(u) as that obtained by $u \in U_{FB}$. Thus, in deterministic control theory, open-loop and feedback controls are equivalent. (The same conclusion holds for closed-loop controls as well). This implies that optimal feedback controls also satisfy the same necessary conditions for optimality given in theorem 3.1. Pontryagin maximum principle does not give controls in the feedback form. For this, one has to proceed via dynamic programming. This leads to a first order partial differential equation, the so-called Hamilton-Jacobi-Bellman equation whose solution, if it exists, immediately gives the optimal control in the feedback form. Thus, we obtain sufficient condition for optimal control as well. Unfortunately, existence of solution of the H-J-B equation needs very strong condition and obtaining the solution is analytically impossible except in trivial situations.

Coming back to the open-loop control law, Theorem 3.1 gives only necessary conditions for optimality and a program satisfying these conditions need not be an optimal program. Sufficient conditions which guarantee that such a program is optimal exist in the literature (see Mangasarian (1966), Kamien and Schwartz (1971) for example). They depend on concavity assumptions of certain functions derived from the Hamiltonian. If one can determine a program satisfying the necessary and sufficient conditions, one obtains an optimal program and the basic optimization problem is solved. In many situations, such explicit determination of a program may be impossible and one may be only interested, for theoretical purposes, in knowing about the existence of an optimal program. There is a huge body of literature dealing with existence questions in optimal control theory and the recent book of Cesari (1983) gives the most exhaustive treatment of the subject matter to date. Sufficiency and existence theorems will not be discussed here.

Let us now consider somewhat more complicated constraints and state necessary conditions for optimality in such situations. They form a subclass of the most general "Hestenes-Bolza" problems, the complete treatment of which may be found in Hestenes (1965). The state now evolves according as

$$\dot{x}(t) = f(t,x(t),\ u(t),b) \tag{3.30}$$

where $x(t) \in \mathbb{R}^n$ is the state, $u(t) \in U_{ad} \subset \mathbb{R}^m$ is the <u>control variable</u> and $b \in B \subset \mathbb{R}^{m'}$ is an additional element, called the <u>control parameter</u>. For $u \in U_{OL}$ and $b \in B$, we want to maximize

$$J(u) = \int_{t_0}^{t_1} \phi(t,x(t),u(t),b)dt + g(b) \tag{3.31}$$

subject to the dynamical constraint (3.30) and an additional set of constraints

$$\chi_i(t,x(t),u(t),b) = 0, \quad i = 1,\dots,l \tag{3.32}$$

with, in general,

$$t_0 = t_0(b), \quad t_1 = t_1(b)$$

$$x(t_0) = x^0(b), \; x(t_1) = x^1(b).$$

We make the following assumptions:

(A1) Allfunctions $\phi, g, f, \chi_i, s, t_0, t_1, x^0, x^1$ are continuously differentiable at points in a set X of the (t,x,u,b)-space;

(A2) Let $X_o \subset X$ satisfy $u(t) \in U_{ad}$, $b \in B$ and $\chi_i(t,x,u,b) = 0$, $i = 1,\dots,l$. Let the matrix $(\frac{\partial\chi}{\partial u} \,\vdots\, \delta_{ij}\chi_j)$ have rank l at each point in X_o, where $\frac{\partial\chi}{\partial u}$ is the Jacobian matrix and δ_{ij} denotes Kronecker delta. (χ is a l-vector with components χ_i, $i = 1,\dots,l$).

<u>Theorem 3.2</u> (Hestenes)

Suppose that (A1)-(A2) hold and $\{x^*(t),u^*(t),b^*;\; t_0 \le t \le t_1\}$ is an optimal program. Then there exist multipliers $p_0, p(t)$ ($n\times 1$-valued function), $q(t)$ ($l\times 1$-valued function), not vanishing simultaneously on $t_0 \le t \le t_1$ and functious L (Lagrangian) and G defined by

$$L(t,x,u,b,p,q) = H(t,x,u,b,p) + q'\chi(t,x,u,b) \tag{3.33}$$

with H defined as

$$H(t,x,u,b,p) = p_o\phi(t,x,u,b) + p'f(t,x,u,b) \tag{3.34}$$

and $G(b) = p_o g(b)$ (3.35)

such that the following necessary conditions hold:

(a) $x^*(t), u^*(t), p(t), q(t)$ satisfy

$$\dot{x}_i^* = \frac{\partial L}{\partial p_i}, \quad \dot{p}_i = -\frac{\partial L}{\partial x_i}, \quad i = 1,\ldots,n \tag{3.36}$$

$$\frac{\partial L^*}{\partial u_i} = 0, \quad i = 1,\ldots,m \tag{3.37}$$

where

$$L^* \equiv L(t,x^*(t),u^*(t),b^*,p(t),q(t)).$$

Moreover,

$$\frac{d}{dt} L^* = \frac{\partial}{\partial t} L^* \tag{3.38}$$

on each interval of continuity of $u^*(t)$ and L^* is continuous on $t_0 \le t \le t_1$.

(b) $$H(t,x^*(t),u^*(t),b^*,p(t)) = \max_{(t,x^*(t),u,b^*) \in X_0} H(t,x^*(t),u,b^*,p(t)) \quad \text{a.e.t} \tag{3.39}$$

(c) The transversality condition holds:

$$-\frac{\partial G}{\partial b_j} + \left[-L_s^* \frac{\partial t_s}{\partial b_j} + \sum_{i=1}^{n} p_i(t_s) \frac{\partial x_i^s}{\partial b_j}\right]_{s=0}^{s=1} = \int_{t_0}^{t_1} \frac{\partial L^*}{\partial b_j}\, dt; \; j = 1,\ldots,m' \tag{3.40}$$

where $L_s^* \equiv L(t,x^*(t_s),u^*(t_s),b^*,p(t_s),q(t_s))$, $s = 0,1$.

In the most general form of Hestenes-Bolza problem, one includes inequality constraints along with equality constraints of the form (3.32) and also integral type inequality and equality constraints. If $t_0 = 0$ and $t_1 = T$, fixed, but $x(0)$ and $x(T)$ are free, one may include them in the control parameters b.

In the original model of optimal economic growth as formulated by Ramsey (1928), one is interested in the existence of a consumption plan which maximizes economic welfare for all generations. For this and many other economic problems, it is natural to consider infinite horizon; i.e., $T = \infty$. In this case, the terminal term in eq^n. (3.14) does not appear. There is a basic problem in the infinite horizon case which arises due to the possible nonconvergence of improper integrals in the criterion. If the criterion takes on an infinite value for a number of admissible controls, it is not possible to compare among them. Halkin (1974) proposed an interesting way out of this dilemma. In his approach, an admissible control is optimal if, for every admissible $\{x(t),u(t)\}$, every positive $T < \infty$ and every $\varepsilon > 0$, there exists $\tau \in [0,\infty]$ such that $\tau \geq T$ and

$$\int_0^\tau \phi(t,x(t),u(t)) \leq \int_0^\tau \phi(t,x^*(t),u^*(t))dt + \varepsilon \tag{3.41}$$

Thus, Halkin's concept of optimality for an infinite horizon problem involves finding a control function which satisfies one of two conditions: either it yields a higher value of the integral than all other paths, or, it is infinitesimally close to other paths. Halkin proved that equations (3.16), (3.17) and (3.19) of Theorem 3.1 are also necessary conditions for the optimality of a program $\{x^*(t),u^*(t);t \geq 0\}$ for infinite horizon problem. Thus, Halkin's result really involves letting $T \to \infty$ in the standard fixed end point problem. The result does not, however, imply that the optimal control for finite T will approach the optimal control for infinite horizon as we let $T \to \infty$.

We close this section with the statement of the corresponding discrete version of the maximum principle given in Theorem 3.1. The system state x_k, $k = 0,1,\ldots,K$ evolves according as

$$x_{k+1} = f_k(x_k,u_k) \tag{3.42}$$

x_0 given

where $x_k \in \mathbf{R}^n$, $u_k \in \mathbf{R}^m$ and is further restricted to lie in $U_{ad} \subset \mathbf{R}^m$ for each k. U_{ad} may be dependent on k. We want to determine an admissible sequence $\{u_k,\ k = 0,1,\ldots,K-1\}$ so as to maximize

$$J(u) = \sum_{k=0}^{K-1} \phi_k(x_{k+1},u_k,x_k). \tag{3.43}$$

The following Theorem is proved in Boltyanski (1978):

<u>Theorem 3.3</u>

Let $\{x_{k+1}^*, u_k^*;\ k = 0,1,\ldots,K-1\}$ denote an optimal discrete program. Then there exists a sequence $\{p_1,\ldots,p_K\}$ of n-dimensional costate vectors and a Hamiltonian function sequence:

$$H_k(x_k,u_k,p_{k+1}) = \phi_k(f_k(x_k,u_k),u_k,x_k) + p'_{k+1}f_k(x_k,u_k) \qquad k = 0,1,\ldots, K-1 \tag{3.44}$$

such that the following necessary conditions hold:

$$x_{k+1}^* = f_k(x_k^*,u_k^*) \tag{3.45}$$

$$x_0^* = x_0$$

$$p_k = \frac{\partial}{\partial x_k} f_k(x_k^*,u_k^*)[p_{k+1} + (\frac{\partial}{\partial x_{k+1}} \phi_k(x_{k+1}^*,u_k^*,x_k^*))'] + \frac{\partial}{\partial x_k} \phi_k(x_{k+1}^*,u_k^*,x_k^*))' \tag{3.46}$$

$$p_K = 0 \tag{3.47}$$

5. APPLICATION TO REGIONAL ALLOCATION OF INVESTMENT

As an application of the maximum principle and also as a background material for our problem in chapter 5, we consider a question in spatial development planning which has a long history in the economic literature. The problem was originally proposed and solved by Rahman (1963). It was extended by Rahman (1966), Intrilligator (1964) and Takayama (1974). They considered two regions, each characterized by a fixed capital-output ratio and each region saving a fixed proportion of its total output. The central planning board (CPB) wants to direct investment of this saving to whichever region it chooses with the objective of maximizing a certain utility function. We study first the Rahman model, make comments on Intrilligator's problem and finally, discuss an extension to nonlinear production function as studied by Fujita (1978). An infinite horizon version of this problem may be found in Pitchford (1977).

In Rahman's formulation, we have two regions with outputs

Y_i $(i = 1,2)$ produced with a fixed capital:output ratio so that we have

$$Y_i = b_i K_i, \qquad i = 1,2 \tag{3.48}$$

with K_i representing capital of region i. With savings propensity of region i assumed constant and equal to s_i, $0 < s_i < 1$, let $\beta(t)$ denote the proportion of total savings invested back into region 1. Then the capital evolution equations of the two regions are given by, with $g_i \equiv b_i s_i$,

$$\dot{K}_1 = \beta(g_1K_1 + g_2K_2), \quad K_1(0) = K_{10} \tag{3.49a}$$

$$\dot{K}_2 = (1-\beta)(g_1K_1 + g_2K_2), \quad K_2(0) = K_{20} \tag{3.49b}$$

K_{10}, K_{20} denoting the initial capital endowment of the two regions. The problem is to choose $\beta(t)$, $0 \le t \le T$, where T is the planning horizon, so as to maximize the total output at the end of the planning period; that is, maximize

$$J(\beta) = Y_1(T) + Y_2(T) \tag{3.50}$$

subject to (3.49a-b) and the constraint $0 \le \beta(t) \le 1$, $0 \le t \le T$. Define the Hamiltonian

$$H \equiv p_1\beta(g_1K_1 + g_2K_2) + p_2(1-\beta)(g_1K_1 + g_2K_2)$$

where the costate variables p_1 and p_2 satisfy

$$\dot{p}_1 = -[\beta(p_1-p_2) + p_2]g_1, \quad p_1(T) = b_1 \tag{3.51a}$$

$$\dot{p}_2 = -[\beta(p_1-p_2) + p_2]g_2, \quad p_2(T) = b_2 \tag{3.51b}$$

Writing H in the form $H = [\beta(p_1-p_2) + p_2](g_1K_1 + g_2K_2)$ it follows from the maximum principle that the optimal β is given by

$$\beta^*(t) = 1 \text{ if } p_1(t) > p_2(t) \tag{3.52a}$$

and

$$\beta^*(t) = 0 \text{ if } p_1(t) < p_2(t) \tag{3.52b}$$

and determination of $\beta^*(t)$ reduces to studying the evolution of $p_1(t)$

and $p_2(t)$ for $0 \le t \le T$. Noting from (3.51a-b) that $\dot{p}_1/\dot{p}_2 = g_1/g_2$, we get

$$p_1(t) = \frac{g_1}{g_2} p_2(t) + \frac{b_1 b_2}{g_2} (s_2 - s_1), \quad p_1(T) = b_1, \quad p_2(T) = b_2 \tag{3.53}$$

or,

$$p_1(t) - p_2(t) = \frac{g_1 - g_2}{g_2} p_2(t) + \frac{b_1 b_2}{g_2} (s_2 - s_1) \tag{3.53'}$$

If $g_1 > g_2$ and $s_2 > s_1$ ($g_1 = g_2$ and $s_2 > s_1$ or $g_1 > g_2$ and $s_2 = s_1$), $p_1(t) > p_2(t)$ so that $\beta^*(t) = 1$ for all t. If, however, $g_1 > g_2$ and $s_1 > s_2$, it is difficult to draw an immediate conclusion and switching of control policy is possible. Since switch occurs when $p_1(t)-p_2(t)$ changes sign, it is obvious that we have to carefully study the case when $p_1(t)-p_2(t) = 0$. Let t^* be a time when this equality occurs and p_2 then takes the value p_2^*. From (3.53'), we see that

$$p_2^* = \frac{s_1 - s_2}{g_1 - g_2} b_1 b_2 \tag{3.54}$$

which is positive, in the case when $g_1 > g_2$ and $s_1 > s_2$. It also readily follows, using $g_i = b_i s_i$, that

$$p_2^* \gtreqless b_2 \text{ according as } b_2 \gtreqless b_1 \tag{3.55}$$

<u>case 1</u> $b_2 > b_1$.

It is clear from eq^{n}s. (3.51a-b) that for optimal β, both $p_1(t)$ and $p_2(t)$ are monotonically decreasing. There is a unique t^* such that $p_2(t^*) = p_2^*$.

(a) $t^* > 0$ (initial time)

There is a switch at time t^* so that the optimal β^* is

$$\beta^*(t) = 1 \quad , \; 0 \le t < t^*$$
$$= 0 \quad , \; t^* \le t \le T$$

(b) $t^* < 0$ (initial time)

There is no switch possible and

$$\beta^*(t) = 0 \quad , \; 0 \le t \le T$$

<u>case 2</u> $b_2 < b_1$.

This case implies $p^* < b_2$ and since $p_2(t)$ decreases monotonically to b_2, this crossing of the magnitude of p_1 and p_2 never occurs. Thus, in this case,

$$\beta^*(t) = 1 \quad , \quad 0 \le t \le T.$$

Intrilligator's problem differed from Rahman's in the optimizing criterion. In this case, one does not maximize the final output but the integrated consumption stream throughout the planning period; thus, the optimizing criterion is

$$J(\beta) = \int_0^T \frac{C}{P} dt$$

where P is the total population and $C = (1-s_1)Y_1+(1-s_2)Y_2$ is the total consumption. The analysis here is somewhat more complicated, although the conclusion is quite similar. See Takayama (1974) for details.

Let us now briefly discuss a more general problem in this framework as formulated and solved by Fujita (1978). Consider a closed economy consisting of n regions. The capital K_i and output Y_i of region i at time t are related by the general production function

$$Y_i = F(K_i) \qquad i = 1,\dots,n \tag{3.56}$$

where $F(\cdot)$ is the same for all regions (in contrast to Rahman's model). Suppose that the savings ratio for the whole economy is a constant s, $0 < s < 1$, so that the total investment available to the CPB is $s(\sum_{i=1}^{n} Y_i(t))$ (this is also in contrast to the Rahman's model where different regions had different savings propensities). Let $\beta_i(t)$ be the proportion of investment allocated back to region i. Then $K_i(t)$ evolves according as

$$\dot{K}_i(t) = \beta_i(t)s \sum_{j=1}^{n} Y_j(t), \quad i = 1,\dots,n \tag{3.57}$$

$$K_i(0) = K_{i0}$$

CPB wants to choose $\beta_i(t)$, $i = 1,\ldots,n$ so as to maximize

$$J = \sum_{i=1}^{n} F(K_i(T) \tag{3.58}$$

the total output at the end of the planning period. We have the obvious constraint

$$\sum_{i=1}^{n} \beta_i(t) = 1, \ \beta_i(t) \geq 0, \ i = 1,\ldots,n; \ 0 \leq t \leq T \tag{3.59}$$

Fujita analyzed all types of production functions F:

(i) decreasing returns to scale; i.e. $F'(K) > 0$ and $F''(K) < 0$ for all $K \geq 0$,

(ii) constant returns to scale; i.e., $F(K) = bK$, $b > 0$, for all K,

(iii) increasing returns to scale; i.e., $F'(K) > 0$ and $F''(K) > 0$ for all $K \geq 0$,

(iv) variable returns to scale; i.e., such that $F'(K) > 0$ for all $K \geq 0$ and there exists a $K^* > 0$ such that $F''(K) > 0$ for $0 \leq K < K^*$, $F''(K^*) = 0$ and $F''(K) < 0$ for $K > K^*$.

With $(a_i)_1^n$ denoting $(a_1,\ldots,a_n)$, the Hamiltonian for the present problem is

$$H((K_i)_1^n, (\beta_i)_1^n, (p_i)_1^n) = (\sum_{i=1}^{n} p_i\beta_i)s(\sum_{i=1}^{n} F(K_i)) \tag{3.60}$$

where $(p_i)_1^n$ denote the costate variables. Let $(\beta_i^*(t))_1^n$ denote the optimal allocation path and $(K_i^*(t))_1^n$ denote the optimal growth path. Then $(p_i(t))_1^n$ satisfies

$$\dot{p}_i(t) = -\frac{\partial H}{\partial K_i}((K_i^*)_1^n, (\beta_i^*)_1^n, (p_i)_1^n) \tag{3.61}$$

$$i = 1,\ldots,n; \quad 0 \leq t \leq T$$

$$p_i(T) = F'(K_i(T)), \ i = 1,\ldots,n \tag{3.62}$$

and the maximum principle asserts that

$$H((K_i^*(t))_1^n, (\beta_i^*(t))_1^n, (p_i(t))_1^n)$$

$$= \max. \{H((K_i^*(t))_1^n, (\beta_i)_1^n, (p_i(t))_1^n | \sum_1^n \beta_i = 1; \ \beta_i \geq 0, \ i = 1,\ldots,n\}$$

$$0 \leq t \leq T \tag{3.63}$$

If we define $p(t) = \max_i . p_i(t)$, $0 \le t \le T$, the Hamiltonian will be maximized if we choose β_i's such that

$$\sum_{i\in Q(t)} \beta_i^*(t) = 1 \text{ where } Q(t) = \{i | p_i(t) = p(t)\} \tag{3.64}$$

Thus all the investment funds of the economy should be allocated among regions which have the highest shadow price for capital at that time. To specify the investment further, one has to study the paths $p_i(t)$ in detail. From eq^n. (3.61),

$$\dot{p}_i(t) = -\left(\sum_{j=1}^{n} \beta_j^*(t) p_j(t)\right) s\, F'(K_i(t)) \tag{3.65}$$

and using optimal β_i's from eq^n. (3.64), we get

$$\dot{p}_i(t) = -\, p(t) s\, F'(K_i(t)) \tag{3.66}$$

This eq^n. along with eq^n. (3.62) leads to the conclusion that $p_i(t) > 0$ and $\dot{p}_i(t) < 0$, $i = 1,\dots,n$, $0 \le t \le T$.

Specializing to the two region economy, we get

$$\dot{K}_i = \beta_i(t) s(F(K_1(t)) + F(K_2(t))), \quad i = 1,2 \tag{3.67}$$

and eq^n. (3.64) specializes to

$$\begin{aligned} &\text{if} \quad p_1(t) > p_2(t), \quad \beta_1(t) = 1 \text{ and } \beta_2(t) = 0 \\ &\text{if} \quad p_1(t) < p_2(t), \quad \beta_1(t) = 0 \text{ and } \beta_2(t) = 1 \\ &\text{if} \quad p_1(t) = p_2(t), \quad \beta_i(t) \text{ is indeterminate, } \beta_1(t) + \beta_2(t) = 1 \end{aligned} \tag{3.68}$$

while eq^n. (3.66) becomes

$$\dot{p}_i(t) = -\max.\,(p_1(t), p_2(t)) sF'(K_i(t)), \quad i = 1,2 \tag{3.69}$$

Detailed analysis of the preceding set of equations led Fujita to the following conclusion: Let K^* be as defined for production functions with variable returns to scale. Define $\psi^*(K)$ by the solution of the following equation:

$$\int_{\psi^*(K)}^{K} \frac{F'(K_2) - F'(K)}{(F(K) + F(K_2))^2}\, dK_2 = 0 \text{ with } K \ge K^* \ge \psi^*(K).$$

Suppose that the planning period T is sufficiently large. Then $\psi^*(K)$ is a single-valued function of K on its domain which is independent of time t, savings ratio s and is symmetric with respect to K_1 and K_2; that is, if $K_2 = \psi^*(K_1)$ for $K_1 \geq K^* \geq K_2$, then $K_1 = \psi^*(K_2)$ for $K_2 \geq K^* \geq K_1$. Furthermore, $\frac{d\psi^*(K)}{dK} < 0$ and denoting $\psi^{*-1}(K')$ to be the inverse function of $\psi^*(K)$, $\frac{d\psi^{*-1}(K')}{dK'} < 0$. The following figure illustrates the function $\psi^*(K)$:

Fig. 3.1

Graphical illustration of $\psi^*(K)$.

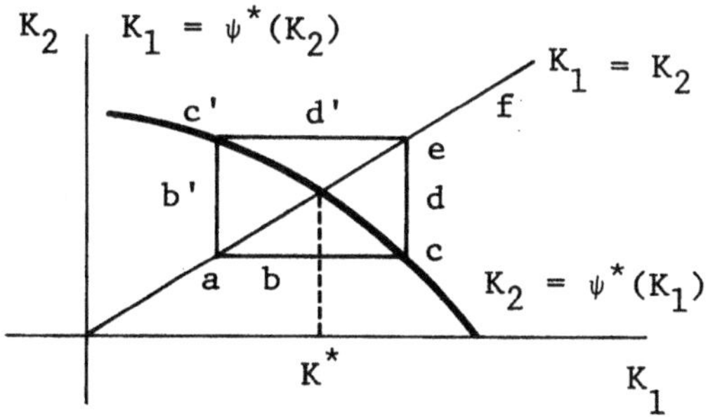

When the initial capital (K_{10}, K_{20}) is at the point b, the optimal capital evolution is given by the path b→c→d→e→f. Thus, in the beginning, all the investment is allocated to the more advanced region 1 until at least upto K^*. On the curve $K_2 = \psi^*(K_1)$ a switch takes place and the whole investment is allocated to the backward region 2 until its capital equals that of region 1. On the other hand, when $K_{10} = K_{20}$, we must decide on one of the regions as our initial favorite one and then the optimal path follows expected pattern with a switch on $K_2 = \psi^*(K_1)$ for $K_1 \geq K^* \geq K_2$ or on $K_1 = \psi^*(K_2)$ for $K_2 \geq K^* \geq K_1$ depending on our initial choice of the favorite region (i.e., the region in which we invest all the available savings). No switching takes place for production functions with only decreasing or increasing returns to scale. One also notices that decreasing returns to scale in production acts as a force for spatial dispersion while increasing returns to scale in production acts as a force for spatial concentration. Similar, although more complicated, results may be obtained for the general n-region problem (See Fujita (1978) for details). In chapter 5, we shall come back to the regional investment allocation problem in the game-theoretic set up.

CHAPTER 4

SOLUTIONS OF HIERARCHICAL DIFFERENTIAL GAMES

1. INTRODUCTION

In this chapter we shall derive necessary conditions for equilibrium solutions of hierarchical differential games formulated in the previous chapter. This will, in general, involve both Stackelberg and Nash differential games. As is obvious from definition, necessary condition for Nash equilibrium solution may be derived by considering component-wise one-person optimization problems. Necessary conditions for Stackelberg equilibrium solution are more complicated, particularly with feedback or more involved information structures. Examples will be given to illustrate the use of the necessary conditions in economic problems.

2. OPEN-LOOP AND FEEDBACK NASH EQUILIBRIA

As mentioned above, one-person optimization results may be directly applied to obtain Nash equilibrium solutions. Necessary conditions for Nash equilibrium can, therefore, be given in a form closely resembling the maximum principle. We consider a N-person differential game described by the state equation

$$\dot{x}(t) = f(t,x(t), u^1(t),\dots, u^N(t)) \tag{4.1}$$

$$x(0) = x_0$$

and with welfare functional for player i:

$$W^i(u^1,\dots,u^N) = \int_0^T \phi^i(t,x(t),u^1(t),\dots,u^N(t))dt + g^i(x(T)) \tag{4.2}$$

Here $u^i(t) \in \mathbb{R}^{m_i}$ denotes the control or decision of player i, which is taken to be piecewise continuous function of time for all i.
We impose the constraint that $u^i(t) \in U^i_{ad}$, $i = 1,\dots,N$, $0 \le t \le T$.

Theorem 4.1

For a N-person differential game formulated above, let $f(t,\cdot,u^1,\ldots,u^N)$, $\phi^i(t,\cdot,u^1,\ldots,u^N)$, $g^i(\cdot)$, $i = 1,\ldots,N$, be continuously differentiable on $\mathbb{R}^n$, $0 \le t \le T$. Let $u^{i*}(t)$, $0 \le t \le T$, $i = 1,\ldots,N$, denote an open-loop Nash equilibrium solution and $\{x^*(t),\ 0 \le t \le T\}$ be the corresponding state trajectory. Then there exist N costate vectors $p^i(t) \in \mathbb{R}^n$, $0 \le t \le T$, and N Hamiltonian functions defined by

$$H^i(t,x,u^1,\ldots,u^N,p^i) = \phi^i(t,x,u^1,\ldots,u^N) + p^{i'}f(t,x,u^1,\ldots,u^N)$$

$$i = 1,\ldots,N, \quad 0 \le t \le T$$

such that the following necessary conditions hold:

$$\dot{x}^*(t) = f(t,x^*(t),u^{1^*}(t),\ldots,u^{N^*}(t)),\ x^*(0) = x_0 \tag{4.3}$$

$$\dot{p}^i_k(t) = -\frac{\partial H^i}{\partial x_k}(t,x^*,u^{1^*},\ldots,u^{N^*}),\ k = 1,\ldots,n,\quad i = 1,\ldots,N \tag{4.4}$$

$$p^i_k(T) = \frac{\partial g^i}{\partial x_k}(x^*(T)),\quad k = 1,\ldots,n,\quad i = 1,\ldots,N \tag{4.5}$$

and for $i = 1,\ldots,N$,

$$H^i(t,x^*(t),u^{1^*}(t),\ldots,u^{i^*}(t),\ldots,u^{N^*}(t),p^i(t))$$

$$= \max_{u^i \in U^i_{ad}} H^i(t,x^*(t),u^{1^*}(t),\ldots,u^{i-1^*}(t),u^i,u^{i+1^*}(t),\ldots,u^{N^*}(t),p^i(t)) \tag{4.6}$$

Proof

In the open-loop case, there is no distinction between strategies and controls and from the definition of Nash equilibrium solution, for $i = 1,\ldots,N$,

$$W^i(u^{1^*},\ldots,u^{i-1^*},u^{i^*},u^{i+1^*},\ldots,u^{N^*}) \ge W^i(u^{1^*},\ldots,u^{i-1^*},u^i,u^{i+1^*},\ldots,u^{N^*})$$

This i-th inequality implies that u^{i^*} maximizes $W^i(u^{1^*},\ldots,u^{i-1^*},u^i,u^{i+1^*},\ldots,u^{N^*})$ over U^i_{ad} subject to the state constraint

$$\dot{x}(t) = f(t,x(t),u^{1^*}(t),\ldots,u^{i-1^*}(t),u^i(t),u^{i+1^*}(t),\ldots,u^{N^*}(t)).$$

But this is a standard optimal control problem for player i since $u^{j^*}(t), j \neq i$, $j = 1,\ldots,N$, are just prescribed functions of time and hence, do not depend upon u^i. The result then follows from the direct application of the continuous-time version of the maximum principle. □

The preceding theorem gives only a set of necessary conditions that any open-loop Nash equilibrium solution must satisfy. To determine a Nash equilibrium solution using theorem 4.1, one has to solve the set of equations (4.3)-(4.6) and from among all the solutions obtained, determine those that satisfy inequalities (3.3) given in definition 3.1. For optimizing criteria which are concave in the control variables, any candidate solution will satisfy the inequalities (3.3) and therefore, qualify for being a Nash equilibrium solution. In fact, for linear dynamics and quadratic criteria functionals which are to be minimized, under the sufficient condition of existence of solution of a coupled set of matrix Riccati equations, the differential game under consideration actually admits a unique Nash open-loop equilibrium solution.

Obtaining necessary conditions for Nash equilibrium solution when one allows for feedback strategies is more complicated. In this case, player i chooses his control or strategy in the form $u^i(t) = \psi^i(t,x(t))$ and this kind of strategy is discussed in detail in Chapter 3, along with the accompanying Nash equilibrium solution concept. Here we give necessary condition for equilibrium in this class of strategies.

Theorem 4.2

We consider a N-person differential game as formulated in the beginning of this chapter. Let $h(t,\cdot), \phi^i(t,\cdot)$, $i = 1,\ldots,N$ be continuously differentiable on $\mathbb{R}^n \times U_{ad}^1 \times \ldots \times U_{ad}^N$, $0 \leq t \leq T$ and $g^i(\cdot)$, $i = 1,\ldots,N$ be continuously differentiable on $\mathbb{R}^n$.

Let $\{\gamma^{i^*}(t,x^*(t)),\ i = 1,\dots,N,\ 0 \le t \le T\}$ denote a feedback Nash equilibrium solution where $x^*(t)$, $0 \le t \le T$, denotes the corresponding state trajectory, such that $\gamma^{i^*}(t,\cdot)$ is continuously differentiable on $\mathbb{R}^n$. Then there exists N costate vectors $p^i(t) \in \mathbb{R}^n$ and N Hamiltonian functions

$$H^i(t,x,u^1,\dots,u^N,p^i) = \phi^i(t,x,u^1,\dots,u^N) + p^{i'} f(t,x,u^1,\dots,u^N)$$
$$i = 1,\dots,N, \quad 0 \le t \le T,$$

such that the following necessary conditions are satisfied:

$$\dot{x}^*(t) = f(t,x^*(t),u^{1^*}(t),\dots u^{N^*}(t));\ x^*(0) = x_0 \tag{4.7}$$

$$\dot{p}^i_k(t) = -\frac{\partial H^i}{\partial x_k}(t,x^*,u^{1^*},\dots,u^{N^*},p^i) - \sum_{\substack{j=1\\ j\neq i}}^{N} \sum_{l=1}^{m_j} \frac{\partial H^i}{\partial u^j_l}(t,x^*,u^{1^*},\dots,u^{N^*},p^i)\frac{\partial \gamma^j_l}{\partial x_k}(t,x^*(t)) \tag{4.8}$$

$$p^i_k(T) = \frac{\partial g^i}{\partial x_k}(x^*(T)) \quad k = 1,\dots,n;\ i = 1,\dots N \tag{4.9}$$

and $u^{i^*}(t) = \gamma^{i^*}(t,x^*(t))$ maximizes

$$H^i(t,x^*(t),\gamma^{1^*}(t,x^*(t)),\dots,\gamma^{i-1^*}(t,x^*(t)),u^i,\gamma^{i+1^*}(t,x^*(t)),\dots,\gamma^{N^*}(t,x^*(t))) \tag{4.10}$$

for $u^i \in U^i_{ad}$.

<u>Proof</u> To determine Nash equilibrium solution, we have to convert criteria W^i's into functions J^i's of $\gamma^i,\dots,\gamma^N$ as explained in the previous chapter. In this transformed representation, $\{\gamma^{i^*}(t,x^*(t)),\ i = 1,\dots N\}$ form an equilibrium solution if

$J^i(\gamma^{1^*},\dots,\gamma^{i-1^*},\gamma^{i^*},\gamma^{i+1^*},\dots,\gamma^{N^*}) \ge J^i(\gamma^{1^*},\dots,\gamma^{i-1^*},\gamma^i,\gamma^{i+1^*},\dots,\gamma^{N^*})$ for all $i = 1,\dots,N$. If we consider one such inequality, say the i-th one, it fixes all players' strategies (except the i-th one) at $\gamma^j = \gamma^{j^*}$ ($j\neq i$, $j = 1,\dots,N$) and constitutes an optimal control problem with feedback strategy for player i.

As explained in the previous chapter, necessary conditions for optimality, the maximum principle, retains the same form for one person optimization with feedback strategy (control). The only complication now is that in the costate equation, the fixed strategies of the other players involve explicitly the current state and this has to be taken into account while calculating the gradient of the Hamiltonian H^i with respect to the state. □

Remark

If we compare the set of equations (4.7)-(4.10) with the set (4.3)-(4.6), we observe that they are identical except for the costate equations which in the former situation has an additional term arising due to the dynamic nature of the feedback information structure. Furthermore, every solution of the set (4.3)-(4.6) also satisfies the equations (4.7)-(4.10), since every such solution is independent of x so that the last term of eq^n. (4.8) becomes zero. Thus open-loop Nash equilibrium is also a feedback Nash equilibrium. This follows also from a general result of Sandell (1974) which, in essence, asserts that if there exists a Nash equilibrium solution strategy with all players restricted to a specified class of strategies, the same equilibrium solution strategy still constitutes a Nash equilibrium solution when all players use enlarged class of strategies. However, open-loop equilibrium solutions need not be the only ones to satisfy equations (4.7)-(4.10). In an important paper, Başar (1977) pointed out that even for linear dynamics and quadratic criteria, one can have a plethora of "informationally nonunique" feedback Nash equilibrium solutions. This feature of feedback Nash equilibrium solutions is discussed in detail in Başar and Olsder (1982). We give below an example, due to Mehlmann (1982), which illustrates this nonuniqueness of feedback Nash equilibria even for linear systems with quadratic criteria.

Example 4.1.

Let us consider a dynamical system evolving according as

$$\dot{x}(t) = A(t)x(t) + B^1(t)u^1(t) + B^2(t)u^2(t) \tag{4.11}$$
$$x(0) = x_0$$

and player i wants to minimize a quadratic functional or equivalently, wants to maximize the negative of that, as given by

$$W^i = -\frac{1}{2}\int_0^T [x'Q^i(t)x + u^{i'}R^{ii}(t)u^i + u^{j'}R^{ij}(t)u^j]dt \tag{4.12}$$
$$-\frac{1}{2}x(T)'Q_f^i\, x(T); \quad i,j = 1,2;\ i \neq j.$$

The matrices $A, B^i, Q^i, R^{ii}, R^{ij}$ are continuous functions of time and we omit the time argument from now on. Q_f^i, Q^i, R^{ii}, R^{ij} are symmetric and we assume that $R^{ii}(t) > 0$ for all t. We assume that $x(t) \in \mathbb{R}^n$ and $u^i(t) \in \mathbb{R}^m$, $i = 1,2$.

Suppose that there exist matrix-valued functions P^1 and P^2 satisfying the following set of coupled Riccati equations, written as a single equation in the augmented form:

$$-\dot{P} = \tilde{A}'P + P\tilde{A} + Q - PSP - PJSPJ - JPSJP + JPJS_0JPJ$$
$$P(T) = P_f \tag{4.13}$$

where

$$A = \begin{bmatrix} A & 0 \\ 0 & A \end{bmatrix}, \quad Q = \begin{bmatrix} Q^1 & 0 \\ 0 & Q^2 \end{bmatrix}, \quad Q_f = \begin{bmatrix} Q_f^1 & 0 \\ 0 & Q_f^2 \end{bmatrix}, \quad P_f = \begin{bmatrix} P_f^1 & 0 \\ 0 & P_f^2 \end{bmatrix}$$

$$S = \begin{bmatrix} B^1(R^{11})^{-1}B^{1'} & 0 \\ 0 & B^2(R^{22})^{-1}B^{2'} \end{bmatrix},$$

$$S_0 = \begin{bmatrix} B^2(R^{22})^{-1}R^{12}(R^{22})^{-1}B^{2'} & 0 \\ 0 & B^1(R^{11})^{-1}R^{21}(R^{11})^{-1}B^{1'} \end{bmatrix}$$

$$P = \begin{bmatrix} P^1 & 0 \\ 0 & P^2 \end{bmatrix}, \quad J = \begin{bmatrix} 0 & I \\ I & 0 \end{bmatrix} \tag{4.14}$$

Suppose that there exist vector-valued functions q^1 and q^2 satisfying the following set of coupled differential equations, written as a single equation in the augmented form:

$$\dot{q} = [PS + PJS - \tilde{A} + J(P + Q')SJ - J(P + Q')JS_0J]q$$
$$+ [JQ'SJP - JQ'JS_0JPJ]y^* \qquad (4.15)$$
$$q(T) = (Q_f - P_f)y^*(T)$$

where

$$Q = \begin{bmatrix} Q^1 & 0 \\ 0 & Q^2 \end{bmatrix}, \quad q = \begin{bmatrix} q^1 \\ q^2 \end{bmatrix}, \quad y^* = \begin{bmatrix} x^* \\ x^* \end{bmatrix} \qquad (4.16)$$

with Q^1, Q^2 arbitrary matrix-valued functions. x may be determined from

$$\dot{y}^* = (\tilde{A} - SP - JSPJ)y^* - (S + JSJ)q; \; y^*(0) = (x_0' \; x_0')' \qquad (4.17)$$

(4.15) and (4.17) constitute a two-point boundary value problem. One can express $y^*(t)$ as

$$y^*(t) = D(t)q(t) + d(t) \qquad (4.17)$$

where

$$\dot{D} = D[\tilde{A} - PS - PJS - J(P + Q')SJ + J(P + Q')JS_0J]$$
$$+ (\tilde{A} - SP - JSP)D + D[JQ'JS_0JPJ - JQ'SJP]D$$
$$- S - JSJ; \quad D(0) = 0 \qquad (4.18)$$

$$\dot{d} = [D(JQ'JS_0JPJ - JQ'SJP) + (\tilde{A} - SP - JSPJ)]d$$
$$d(0) = (x_0' \; x_0')' \qquad (4.19)$$

Mehlmann (1982) has shown that any strategy pair of the affine form

$$\gamma^{i*}(t,x) = -(R^{ii})^{-1}B^{i'}[(P^i + Q^i)x - Q^ix^* + q^i], \; i = 1,2 \qquad (4.20)$$

is a feedback Nash equilibrium solution if the following conditions hold:

(C1) There exist unique and continuously differentiable solutions P(t) of (4.13) and D(t) of (4.18) and $Q^1(t)$, $Q^2(t)$ are taken to be continuously differentiable matrix-valued functions;

(C2) The matrix $[I - D(T)(Q_f - P_f)]^{-1}$ exists;

(C3) The matrices Q_f and $Q + J(P + Q')JS_0(P + Q')J$ are positive semidefinite for $0 \le t \le T$.

Thus, although the corresponding open-loop problem has a unique Nash equilibrium solution, the situation in feedback strategies is entirely different. We have uncountable number of Nash equilibrium solution strategies of the affine form and it may even be possible that there exist nonlinear Nash equilibrium strategies. However, in the two extreme (degenerate) cases of team problem and zero-sum differential games, one gets again unique feedback solutions, a familiar result in linear-quadratic control problems.

We end this section with an illustration of the use of the necessary conditions developed above. We do this by applying our results to the Cournot duopoly model in a dynamic context. This was studied by Simaan and Takayama (1978) and is an extension of static Cournot duopoly model studied in example 2.1.

Example 4.2

Let us consider two firms, denoted 1 and 2, producing outputs $q^1(t)$ and $q^2(t)$ of a product at time t, respectively. Price is, of course, determined by the supply and demand for the commodity. Total supply is clearly $q(t) = q^1(t) + q^2(t)$ and suppose that the price function evolution is given by

$$\dot{p}(t) = G[h(q(t)) - p(t)] \tag{4.21}$$

$$p(0) = p_0; \quad 0 \le t \le T$$

where $[0,T]$ is the planning period. We assume that $G[\cdot]$ is a monotonically increasing function of its argument; more specifically,

$$G[0] = 0; \quad \frac{dG[x]}{dx} > 0 \qquad \text{for all } x.$$

G is interpreted as the speed of adjustment of the price function. We also assume that $h(\cdot)$ is a monotonically decreasing function of its argument; that is, $\frac{dh(y)}{dy} < 0$, implying that an increase in total supply of the product will cause a decrease in its price and vice versa.

The outputs $q^1(t)$ and $q^2(t)$ of the two firms F1 and F2 are at their disposal and they are free to decide on their respective outputs so as to maximize their individual profits. Let $c^i(q^i(t))$ denote the cost of production per unit time at time t for Fi, where $c^i(\cdot)$ is, in general, a convex function of its argument, $i = 1,2$. With price $p(t)$ of the product, the unit time profit of Fi at time t is clearly $p(t)q^i(t) - c^i(q^i(t))$, so that the accumulated profit for Fi is

$$\Pi^i(q^1,q^2) = \int_0^T \exp.(-\rho_i t)[p(t)q^i(t) - c^i(q^i(t))]dt; \; i = 1,2 \quad (4.22)$$

with ρ_i denoting a discount factor.

We use theorem 4.2 to determine a Nash equilibrium solution because it can be easily specialized to obtain open-loop Nash equilibrium solution as well. In practice, one always has to assume an upper limit to the production capacity of Fi, which we denote by Q^i. This acts as a control constraint for our problem. Strategy of Fi at time t has the form $q^i(t) = \gamma^i(t,p(t))$ where γ^i has to be chosen so that eq^n. (4.21) has a unique solution and the corresponding $q^i(t)$ satisfies

$$0 \leq q^i(t) \leq Q^i, \; i = 1,2; \quad 0 \leq t \leq T.$$

Let us first work with the Hamiltonian associated with F1 given by

$$\tilde{H}(t,p,q^1,q^2,\mu^1) = \exp.(-\rho_1 t)\{p(t)\gamma^1(t,p(t)) - c^1(\gamma^1(t,p(t)))\} + \mu^1(t)f(p(t),\gamma^1(t,p(t)) + \gamma^2(t,p(t)) \quad (4.23)$$

where

$$f(p(t),q(t)) \equiv G[h(q(t))-p(t)] \quad (4.24)$$

From theorem 4.2, the necessary conditions are given by

$$\frac{dp^*(t)}{dt} = f(p^*(t),\ q^*(t)) \quad , \quad p^*(0) = p_0 \tag{4.25}$$

$$\dot{\mu}^1(t) = -\exp.(-\rho_1 t)[q^1 + p\,\frac{\partial\gamma^1}{\partial p} - \frac{dc^1}{dq^1}\,\frac{\partial\gamma^1}{\partial p}]$$

$$-\mu^1[\frac{\partial f}{\partial p} + \frac{\partial f}{\partial q^1}\,\frac{\partial\gamma^1}{\partial p} + \frac{\partial f}{\partial q^2}\,\frac{\partial\gamma^2}{\partial p}] \ ,\ \mu^1(T) = 0 \tag{4.26}$$

and

$$\tilde{H}(t,p^*(t),\gamma^{1^*}(t,p^*(t)),\ \gamma^{2*}(t,p^*(t)),\ \mu^1(t))$$

$$= \max_{0\leq q^1 \leq Q^1}.\ H^1(t,p^*(t),q^1,\gamma^{2^*}(t,p^*(t)),\mu^1(t)) \tag{4.27}$$

We now use a technique standard in optimization problems with a discounting factor. Let $\lambda^1 = \exp.(\rho_1 t)\mu^1$ and define

$$H^1(t,p(t),q^1(t),q^2(t),\lambda^1(t)) = \exp.(\rho_1 t)\tilde{H}^1(t,p(t),q^1(t),q^2(t),\mu^1(t)) \tag{4.28}$$

Eq^n. (4.26) then reduces to

$$\dot{\lambda}^1 = \rho_1\lambda^1 - q^1 - (p- \frac{dc^1}{dq^1} + \lambda^1\,\frac{\partial f}{\partial q^1})\,\frac{\partial\gamma^1}{\partial p}\bigg|_*$$

$$-\lambda^1(\frac{\partial f}{\partial p} + \frac{\partial f}{\partial q^2}\,\frac{\partial\gamma^2}{\partial p})\bigg|_* \ ,\ \lambda^1(T) = 0 \tag{4.29}$$

where * indicates that we evaluate these quantities on the optimal path. To obtain the optimal $q^{1*} = \gamma^{1^*}(t,p^*(t))$, we have to maximize H^1 with respect to q^1 for $0 \leq q^1 \leq Q^1$. The elementary necessary condition for this is given by

$$\frac{dH^1}{dq^1} \begin{cases} = 0 & \text{for} \quad 0 < q^1 < Q^1 \\ < 0 & \text{for} \quad q^1 = 0 \\ > 0 & \text{for} \quad q^1 = Q^1 \end{cases}$$

which, in our present context, may be reexpressed as

$$p - \frac{dc^1}{dq^1} + \lambda^1\frac{\partial f}{\partial q^1} \begin{cases} = 0 & \text{for} \quad 0 < \gamma^{1^*}(t,p^*(t)) < Q^1 \\ < 0 & \text{for} \quad \gamma^{1^*}(t,p^*(t)) = 0 \end{cases} \tag{4.30}$$

$> 0 \quad \text{for} \quad \gamma^{1^*}(t,p^*(t)) = Q^1$

Since, on the boundary, γ^1 is independent of p, condition (4.30) implies that, at the optimum,

$$(p- \frac{dc^1}{dq^1} + \lambda^1 \frac{\partial f}{\partial q^1}) \frac{\partial \gamma^1}{\partial p} = 0 \tag{4.31}$$

and eq^n. (4.29) becomes

$$\dot{\lambda}^1 = \rho_1\lambda^1 - \gamma^{1^*}(t,p^*(t)) - \lambda^1(\frac{\partial f}{\partial p} + \frac{\partial f}{\partial q^2}\frac{\partial \gamma^2}{\partial p})\Big|_* \tag{4.29'}$$

Similar results are readily obtained for F2. We, therefore, obtain the following result:

Any Nash equilibrium solution $q^{1^*}(t) = \gamma^{1^*}(t,p^*(t))$ and $q^{2^*}(t) = \gamma^{2^*}(t,p^*(t))$ must satisfy the following necessary conditions:

$$\dot{p}^*(t) = f(p^*(t), \gamma^{1^*}(t,p^*(t)), \gamma^{2^*}(t,p^*(t))), \; p^*(0) = p_0 \tag{4.32}$$

$$\dot{\lambda}^1(t) = \rho_1\lambda^1(t) - \gamma^{1^*}(t,p^*(t)) - \lambda^1(t) (\frac{\partial f}{\partial p} + \frac{\partial f}{\partial q^2}\frac{\partial \gamma^2}{\partial p})\Big|_* \tag{4.33a}$$

$$\dot{\lambda}^2(t) = \rho_2\lambda^2(t) - \gamma^{2^*}(t,p^*(t)) - \lambda^2(t) (\frac{\partial f}{\partial p} + \frac{\partial f}{\partial q^1}\frac{\partial \gamma^1}{\partial p})\Big|_* \tag{4.33b}$$

$$(p - \frac{dc^i}{dq^i} + \lambda^i\frac{\partial f}{\partial q^i})\Big|_* \begin{cases} = 0 & \text{for } 0 < \gamma^{i^*}(t,p^*(t)) < Q^i \\ < 0 & \text{for } \gamma^{i^*}(t,p^*(t)) = 0 \\ > 0 & \text{for } \gamma^{i^*}(t,p^*(t)) = Q^i \end{cases} \tag{4.34}$$

The solution may show any combination of nine possibilities given in eq^n. (4.34), each being true for a subinterval of [0,T]. It is generally impossible to solve analytically the partial differential equations (4.32), (4.33a), (4.33b). In the following, we specialize to the case of linear price evolution equation and quadratic cost function.

Suppose that p(t) satisfies a linear differential equation of the form

$$\dot{p}(t) = a[c-b(q^1(t)+q^2(t))-p(t)] \tag{4.35}$$

and the cost functions have the form

$$c^i(x) = \frac{1}{2}\alpha^i x^2 \quad , \quad i = 1,2 \tag{4.36}$$

where a,b,c,α^1,α^2 are positive constants. Assume also that $\rho_1 = \rho_2 = \rho$. Then the maximizing criterion (profit) for firm i has the form

$$\Pi^i(q^1,q^2) = \int_0^T \exp.(-\rho t)[p(t)q^i(t) - \tfrac{1}{2}\alpha^i q^i(t)^2]dt \qquad (4.37)$$

To ensure that p(t) always remains positive, we impose the constraint that $q^1 + q^2 \leq \frac{c}{b}$. Defining $\tilde{p}(t) = c - p(t)$, it follows from (4.35) that

$$\dot{\tilde{p}}(t) \geq - a\,\tilde{p}(t)$$

implying that $\tilde{p}(t) \geq 0$ for all t or, equivalently, $p(t) \leq c$ for all t. In the (p,q) plane, we are therefore only interested in the region $0 \leq p \leq c$ and $0 \leq q \leq \frac{c}{b}$. Equations (4.32)-(4.34) specialize now to the following set of necessary conditions:

$$\dot{p}^*(t) = a[c-b(\gamma^{1^*}(t,p^*(t)) + \gamma^{2^*}(t,p^*(t)))-p(t)],\; p^*(0) = p_0 \qquad (4.38)$$

$$\dot{\lambda}^1(t) = [\rho+a+ab \frac{\partial\gamma^{2^*}(t,p^*(t))}{\partial p}]\,\lambda^1(t) - \gamma^{1^*}(t,p^*(t)) \qquad (4.39a)$$

$$\dot{\lambda}^2(t) = [\rho+a+ab \frac{\partial\gamma^{1^*}(t,p^*(t))}{\partial p}]\,\lambda^2(t) - \gamma^{2^*}(t,p^*(t)) \qquad (4.39b)$$

$$\lambda^1(T) = \lambda^2(T) = 0$$

and for i = 1,2,

$$p^*(t)-\alpha^i\gamma^{i^*}(t,p^*(t)) - ab\lambda^i(t) \begin{cases} = 0 & \text{for } 0 < \gamma^i(t,p^*(t)) < Q^i \\ < 0 & \text{for } \gamma^i(t,p^*(t)) = 0 \\ > 0 & \text{for } \gamma^i(t,p^*(t)) = Q^i \end{cases} \qquad (4.40)$$

The solution of the above system of equations may be described by the nine possibilities described below:

Possibility	$\gamma^{1^*}(t,p^*(t))$	$\gamma^{2^*}(t,p^*(t))$
P1	0	0
P2	0	≥ 0
P3	≥ 0	0
P4	0	Q^2
P5	Q^1	0
P6	≥ 0	Q^2
P7	Q^1	≥ 0
P8	Q^1	Q^2
P9	≥ 0	≥ 0

TABLE 4.1 Possibilities of solution of the duopoly market

All these nine cases have been analyzed in detail in Simaan and Takayama (1978). Only case P9 is a real duopolist situation in the sense that both firms are actively engaged in a competitive market and it is only in this situation that open-loop and feedback strategies lead to two different solutions. We do not pursue this here any further.

3. OPEN-LOOP STACKELBERG SOLUTION FOR 2-PERSON DIFFERENTIAL GAMES

In this section, we give necessary conditions for open-loop Stackelberg equilibrium solution in 2-person differential games for a fixed time duration $[0,T]$. The state $x(t) \in \mathbb{R}^n$ is described by

$$\dot{x}(t) = f(t,x(t),\ u^1(t),\ u^2(t)) \tag{4.41}$$
$$x(0) = x_0,\ \text{given}$$

where $u^i(t) \in U^i_{ad} \subset \mathbb{R}^{m_i}$, $i = 1,2$, for each t. Player i wants to maximize his own welfare functional

$$W^i(u^1,u^2) = \int_0^T \phi^i(t,x,u^1,u^2)dt + g^i(x(T)) \tag{4.42}$$

and we assume that player 1 acts as the leader. Since we are concerned with open-loop strategies, we can define Stackelberg equilibrium solution concept directly in terms of the control functions instead of representing the criteria in terms of strategies and defining equilibrium solution via that form of representation. Let U^i denote the set of admissible controls for player i; that is, all $\mathbb{R}^{m_i}$-valued piecewise-continuous functions on $[0,T]$ whose values at each t are restricted to lie in a set $U^i_{ad} \subset \mathbb{R}^{m_i}$, $i = 1,2$. In Stackelberg game, the follower responds to the announced strategy of the leader. Thus, for fixed $u^1 \in U^1$, the follower, player 2, is faced with the following problem: determine

$$\max_{u^2 \in U^2} W^2(u^1,u^2)\quad,\quad u^1 \text{ fixed} \tag{4.43}$$

subject to

$$\dot{x}(t) = f(t,x(t),u^1(t),u^2(t)) \tag{4.44}$$
$$x(0) = x_0,\ \text{given}$$

This is a standard optimal control problem. In fact, under the following sufficient conditions:

(c1) W^i are continuous on $U^1 \times U^2$

(c2) $W^2(u^1,\cdot)$ is strictly concave on U^2 for all $u^1 \in U^1$

(c3) U^i_{ad} are compact subsets of $\mathbb{R}^{m_i}$

the above control problem for the follower has a unique optimal solution. Anyway, the following lemma gives necessary conditions for optimality of the follower's optimization problem:

Lemma 4.1

Assume that $f(t,\cdot,u^1,u^2)$, $\phi^2(t,\cdot,u^1,u^2)$ and $g^2(\cdot)$ are continuously differentiable on $\mathbb{R}^n$. Under the sufficient conditions mentioned above, to any announced strategy $u^1(\cdot)$ of the leader, there is a unique optimal response, denoted $\bar{u}^2(\cdot)$, of the follower and suppose that $\bar{x}(\cdot)$ denote the corresponding optimal state trajectory. Then there exists a costate vector $p(t) \in \mathbb{R}^n$, $0 \le t \le T$, and a Hamiltonian

$$H^2(t,x,u^1,u^2,p) = \phi^2(t,x,u^1,u^2) + p'f(t,x,u^1,u^2) \tag{4.45}$$

such that the following conditions hold:

$$\dot{x} = f(t,\bar{x}(t),u^1(t),\bar{u}^2(t)) \tag{4.46}$$

$$\dot{p}_i(t) = -\frac{\partial \phi^2}{\partial x_i}(t,\bar{x},u^1,\bar{u}^2) - \sum_{j=1}^{n} p_j(t)\,\frac{\partial f_i}{\partial x_j}(t,\bar{x},u^1,\bar{u}^2) \tag{4.47}$$

$$p_i(T) = \frac{\partial g^2}{\partial x_i}(\bar{x}(T)) \tag{4.48}$$

and for a.e. $t \in [0,T]$,

$$H^2(t,\bar{x}(t),u^1(t),\bar{u}^2(t),p(t)) = \max_{u^2 \in U^2_{ad}} H^2(t,\bar{x}(t),u^1(t),u^2,p(t)) \tag{4.49}$$

Proof Necessary conditions follow directly from Pontryagin's maximum principle. □

To obtain the optimal control (decision) of the leader, we impose some further conditions:

(c4) $f(t,x(t),u^1(t),\cdot)$ is continuously differentiable on U^2_{ad};

(c5) $\phi^2(t,x(t),u^1(t),\cdot)$ is continuously differentiable on U^2_{ad};

(c6) $\bar{u}^2$ in lemma 4.1 is an interior point solution of the maximizing of the Hamiltonian for every $u^1(\cdot) \in U^1$.

Then clearly eq^n. (4.48) will imply that

$$\frac{\partial H^2}{\partial u_j^2}(t,x(t),u^1(t),\bar{u}^2(t),p(t)) = 0 \ , \quad j = 1,\dots,m_2 \tag{4.50}$$

To obtain the Stackelberg optimal control (strategy) for the leader, we have to maximize $W^1(u^1,u^2)$ with respect to $u^1(\cdot) \in U^1$ and $u^2(\cdot) \in U^2$ subject to the constraints (4.46), (4.47), (4.48) and (4.50). More explicitly, player 1, the leader, is faced with the following problem:

determine $\max\limits_{\substack{u^1(\cdot)\in U^1 \\ u^2(\cdot)\ U^2}} \quad W^1(u^1,u^2)$

subject to

$$\dot{x}(t) = f(t,x(t),u^1(t),u^2(t)), \ x(0) = x_0 \tag{4.46'}$$

$$\dot{p}(t) = F(t,x(t),u^1(t),u^2(t),p(t)) \tag{4.47'}$$

$$p(T) = p_T \ , \ \text{given} \tag{4.48'}$$

$$\frac{\partial H^2}{\partial u_j^2}(t,x(t),u^1(t),u^2(t),p(t)) = 0 \text{ for all } t, \quad j = 1,\dots,m_2 \tag{4.50}$$

where

$$F(t,x(t),u^1(t),u^2(t),p(t)) = -(\nabla_x\phi^2(t,x(t),u^1(t),u^2(t)))' \\ -\dot{p}(t)'\nabla_x f(t,x(t),u^1(t),u^2(t)) \tag{4.51}$$

and

$$p_T = (\nabla_x g^2(x(T)))' \tag{4.52}$$

where ∇_x is clearly the row vector denoting gradient with respect to x. While deriving necessary conditions for optimality of the optimization problem above, we note that the dynamical equations for the states for this problem, viz., x(t) and p(t), involve boundary conditions at both the end points. The necessary condition for optimality of the control (strategy) for the leader is given by the following:

Theorem 4.3

In addition to conditions (c1)-(c6) and assumptions of lemma 4.1, suppose that

(i) $f(t,x(t),\cdot,u^2(t))$, $\phi^2(t,x(t),\cdot,u^2(t))$ are continuously differentiable on U^1_{ad},

(ii) $\phi^1(t,\cdot,\cdot,\cdot)$ is continuously differentiable on $\mathbb{R}^n \times U^1_{ad} \times U^2_{ad}$,

(iii) $f(t,\cdot,u^1(t),\cdot)$ and $\phi^2(t,\cdot,u^1(t),\cdot)$ are twice continuously differentiable on $\mathbb{R}^n \times U^2_{ad}$.

Let $u^{1^*}(\cdot)$ be an open-loop Stackelberg equilibrium strategy (control) for the leader with $u^{1^*}(t)$ belonging to the interior of U^1_{ad} for all t. Let $u^{2^*}(\cdot)$ be the corresponding optimal strategy (control) for the follower as determined by lemma 4.1, $p^*(t)$ and $x^*(t)$, $0 \le t \le T$, the corresponding costate vector and optimal state, respectively. Then there exists a Lagrangian (generalized Hamiltonian)

$$\begin{aligned} &L(t,x(t),p(t),u^1(t),u^2(t),\lambda^1(t),\lambda^2(t),\gamma(t)) \\ &= \phi^1(t,x(t),u^1(t),u^2(t)) + \lambda^1(t)'f(t,x(t),u^1(t),u^2(t)) \\ &+ \lambda^2(t)'F(t,x(t),u^1(t),u^2(t),p(t)) \\ &+ \sum_{j=1}^{m_2} \gamma_j(t) \frac{\partial H^2}{\partial u_j^2}(t,x(t),u^1(t),u^2(t),p(t)) \end{aligned} \tag{4.53}$$

such that the following relations hold:

$$\dot{x}^*(t) = f(t,x^*(t), u^{1^*}(t),u^{2^*}(t)),\ x^*(0) = x_0 \tag{4.54}$$

$$\dot{p}^*(t) = F(t,x^*(t), u^{1^*}(t),u^{2^*}(t),p^*(T)),\ p^*(T) = p_T \tag{4.55}$$

$$\dot{\lambda}_k^1(t) = -\frac{\partial L}{\partial x_k}(t,x^*(t), p^*(t), u^{1^*}(t),u^{2^*}(t),\lambda^1(t),\lambda^2(t),\gamma(t)) \tag{4.56}$$

$$\lambda_k^1(T) = \frac{\partial g^1}{\partial x_k}(x(T)) - \sum_{j=1}^{n} \lambda_j^2(T) \frac{\partial^2 g^2}{\partial x_j \partial x_k}(x(T)) \tag{4.57}$$

$$\dot{\lambda}_k^2(t) = -\frac{\partial L}{\partial p_k}(t,x^*(t),p^*(t),u^{1^*}(t),u^{2^*}(t),\lambda^1(t),\lambda^2(t),\gamma(t)) \tag{4.58}$$

$$\lambda_k^2(0) = 0, \quad k = 1,\ldots,n \tag{4.59}$$

$$\nabla_{u^1} L(t,x^*(t),p^*(t),u^{1^*}(t),u^{2^*}(t),\lambda^1(t),\lambda^2(t),\gamma(t)) = 0 \tag{4.60}$$

$$\nabla_{u^2} L(t,x^*(t),p^*(t),u^{1^*}(t),u^{2^*}(t),\lambda^1(t),\lambda^2(t),\gamma(t)) = 0 \tag{4.61}$$

Proof As discussed above, the leader (player 1) is faced with an optimal control problem of maximizing $W^1(u^1,u^2)$ over $U^1 \times U^2$ subject to $eq^n s$. (4.46')-(4.48') and (4.50'). To obtain the necessary conditions, we apply theorem 3.2 where we take $(x(t)'p(t)')'$ as the system state at time t and consider the vector $(p(0)'x(T)')'$ as the control parameters b of theorem 3.2 while condition (4.50'), i.e., $\frac{\partial H^2}{\partial u_j^2}(t,x(t),u^1(t),u^2(t),p(t)) = 0$ for $j = 1,\ldots,m_2$, acts as constraints. In order that $(u^{1^*}(t),u^{2^*}(t))$ be optimal controls (decisions) with $(x^*(t),p^*(t))$ the corresponding state of the augmented system, the following necessary conditions must hold:

(a) there exist multiplier vectors $\lambda^1(t)$, $\lambda^2(t)$, $\gamma(t)$, $0 \le t \le T$ such that $x^*(t)$, $p^*(t)$, $u^{1^*}(t)$, $u^{2^*}(t)$, $\lambda^1(t)$, $\lambda^2(t)$, $\gamma(t)$ satisfy the following differential equations:
with $z^*(t) = (x^*(t)\ p^*(t)')'$ and $\lambda(t) = (\lambda^1(t)'\ \lambda^2(t)')'$,

$$\dot{z}^* = \nabla_\lambda L \quad , \quad \dot{\lambda} = -\nabla_{z^*} L$$

i.e.,

$$\dot{x}^*(t) = \frac{\partial L}{\partial \lambda^1} \quad , \quad \dot{p}^*(t) = \frac{\partial L}{\partial \lambda^2} \tag{4.62}$$

$$\dot{\lambda}^1(t) = -\frac{\partial L}{\partial x^*} \quad , \quad \dot{\lambda}^2(t) = -\frac{\partial L}{\partial p^*} \tag{4.63}$$

Furthermore,

$$\nabla_{u^1} L^* = \nabla_{u^2} L^* = 0 \tag{4.64}$$

where L^* is defined by

$$L^* = L(t,x^*(t),p^*(t),u^{1^*}(t),u^{2^*}(t),\lambda^1(t),\lambda^2(t),\gamma(t)) \tag{4.65}$$

with L defined by eq^n. (4.53).

(b) The following transversality condition holds:

$$-\frac{\partial G}{\partial b_j} + [-L_s^* \frac{\partial t_s}{\partial b_j} + \sum_{i=1}^{2n} \lambda_i(t_s) \frac{\partial z_i^{*s}}{\partial b_j}] \Big|_{s=0}^{s=1} = \int_0^T \frac{\partial L^*}{\partial b_j} dt, \quad j = 1,..,2n$$

where $t_0 = 0$, $t_1 = T$ and $G(b) = g(b)$ $(p_0 = 1)$.

Since ϕ, f_i's , χ_i's (of Theorem 3.2) in our specific case here do not depend explicitly on b, so that L does not contain b explicitly, the transversality condition is identically zero. Furthermore, t_0 (=0) and t_1 (=T) being fixed, the transversality condition becomes

$$-\nabla_b g(b) + \lambda(T)'\nabla_b z^*(T) - \lambda(0)' \nabla_b z^*(0) = 0.$$

Separating the x and p parts give

$$- \nabla_x g^1(x(T)) + \lambda^1(T)' + \lambda^2(T)'\nabla_x p(T) = 0$$

$$\lambda^2(0)' = 0$$

yielding the desired boundary conditions on noting that p(T) satisfies eq^n. (4.52) □

4. LINEAR-QUADRATIC STACKELBERG GAMES

Linear-quadratic Stackelberg games may be solved by applying results of the previous section. We solve these games by a different technique in this section where we use a function space approach. We follow Simaan and Cruz (1973) who first explicitly solved the linear-quadratic Stackelberg game using the Hilbert space approach. The advantage of this technique is that it can be readily extended to stochastic games (see Bagchi and Başar (1981), for example). In intertemporal optimization problems arising in economics, linear-quadratic model is inappropriate in most situations. In short term stabilization policies, however, the linear-quadratic model seems realistic and has been studied by Pindyck (1977)) and others.

The state dynamics of the game now is given by the linear differential equation

$$\dot{x} = Ax + B^1u^1 + B^2u^2 \tag{4.66}$$

$$x(0) = x_0 \text{ , given}$$

where the system matrices A, B^1 and B^2 may be time dependent. The quadratic performance criteria which are to be minimized are given by

$$J^1(u^1,u^2) = \frac{1}{2} \int_0^T [x(t)'Q^1(t)x(t)+u^{1'}R^{11}(t)u^1+u^{2'}R^{12}u^2]dt + \frac{1}{2} x(T)'Q_f^1x(T) \tag{4.67a}$$

$$J^2(u^1,u^2) = \frac{1}{2} \int_0^T [x(t)'Q^2(t)x(t)+u^{1'}R^{21}(t)u^1+u^{2'}R^{22}(t)u^2]dt + \frac{1}{2} x(T)'Q_f^2x(T) \tag{4.67b}$$

Here $x(t) \in \mathbb{R}^n$, $u^i(t) \in \mathbb{R}^{m_i}$ (Lebesgue) square integrable functions on $[0,T]$, the matrices appearing in eq^{n}s. (4.67a)-(4.67b) are symmetric with $Q_f^i \geq 0$, $Q^i(t) \geq 0$, $R^{ii}(t) > 0$ and $R^{ij}(t) \geq 0$, $i \neq j$, $i,j = 1,2$ (Here ≥ 0 indicates nonnegative definiteness while > 0 indicates positive definiteness). $L_2^n[0,T]$ denotes the space of all $\mathbb{R}^n$-valued measurable functions f on $[0,T]$ which satisfy $\int_0^T \|f(t)\|^2 dt < \infty$. Define a product Hilbert space $H = L_2^n [0,T] \times \mathbb{R}^n$ with the inner product*. Eq^n. (4.66) can be explicitly solved as

$$x(t) = \Phi(t,0)x_0 + \int_0^t \Phi(t,\tau)B^1(\tau)u^1(\tau)d\tau + \int_0^t \Phi(t,\tau)B^2(\tau)u^2(\tau)d\tau \tag{4.68}$$

where $\Phi(t,s)$ is the state transition matrix for $\dot{x} = A(t)x$, so that evaluating $x(t)$ at t and T, we can write eq^n. (4.68) as an element in H satisfying

$$\tilde{x} = \Phi\tilde{x}_0 + L^1u^1 + L^2u^2 \tag{4.69}$$

where

$$\Phi = \begin{bmatrix} \Phi(t,0) & 0 \\ 0 & \Phi(T,0) \end{bmatrix}, \quad \tilde{x} = \begin{bmatrix} x(t) \\ x(T) \end{bmatrix} \in H, \quad \tilde{x}_0 = \begin{bmatrix} x_0 \\ x_0 \end{bmatrix} \in H \tag{4.70}$$

L^i, $i = 1,2$, are linear operators mapping $L_2^{m_i} [0,T]$ into H defined by

---

* $\left[\langle [\begin{smallmatrix} f \\ x \end{smallmatrix}], [\begin{smallmatrix} g \\ y \end{smallmatrix}] \rangle = \int_0^T f(t)'g(t)dt + x'y \; ; \; [\begin{smallmatrix} f \\ x \end{smallmatrix}], [\begin{smallmatrix} g \\ y \end{smallmatrix}] \in H \right]$

$$\mathcal{L}^i u^i = \begin{bmatrix} \int_0^t \Phi(t,\tau)B^i(\tau)u^i(\tau)d\tau \\ \\ \int_0^T \Phi(t,\tau)B^i(\tau)u^i(\tau)d\tau \end{bmatrix}, \quad i = 1,2 \tag{4.71}$$

We have to convert the performance criteria into quadratic functionals on H. Defining

$$\tilde{Q}^i\tilde{x} = \begin{bmatrix} Q^i(t)x(t) \\ Q^i_f x(T) \end{bmatrix}, \quad i = 1,2$$

the performance criteria may be reexpressed as

$$J^1(u^1,u^2) = \tfrac{1}{2}[<\tilde{x},\tilde{Q}^1\tilde{x}>+<u^1,R^{11}u^1>+<u^2,R^{12}u^2>] \tag{4.72a}$$

$$J^2(u^1,u^2) = \tfrac{1}{2}[<\tilde{x},\tilde{Q}^2\tilde{x}>+<u^1,R^{21}u^1>+<u^2,R^{22}u^2>] \tag{4.72b}$$

where $<\cdot,\cdot>$ stands for inner products in appropriate Hilbert spaces. Note that we use the same symbol to denote inner products in different Hilbert spaces. We assume, as before, that player 1 acts as the leader. If u^1 is fixed, player 2 can calculate his optimal strategy by minimizing $J^2(u^1,u^2)$ for fixed u^1. Substituting for $\tilde{x}$ from (4.69) in (4.72b), and using '*' to denote adjoint of an operator, we can write

$$J^2(u^1,u^2) = \tfrac{1}{2}[<u^2,(R^{22}+\mathcal{L}^{2^*}\tilde{Q}^2\mathcal{L}^2)u^2> + 2<u^2,\mathcal{L}^{2^*}\tilde{Q}^2(\Phi x_0+\mathcal{L}^1u^1)> + J_{20}] \tag{4.73}$$

where $J_{20} = [<\Phi\tilde{x}_0 + \mathcal{L}^1u^1,\tilde{Q}^1(\Phi\tilde{x}_0+\mathcal{L}^1u^1)>+<u^1,R^{21}u^1>]$.

To determine the minimizing u^2, we can set the gradient of $J^2(u^1,u^2)$ w.r.t. u^2 to zero, yielding

$$(R^{22}+\mathcal{L}^{2^*}\tilde{Q}^2\mathcal{L}^2)u^2 + \mathcal{L}^{2^*}\tilde{Q}^2(\Phi\tilde{x}_0+\mathcal{L}^1u^1) = 0.$$

Our assumption that $R^{22} > 0$, $Q^2_f \geq 0$ and $Q^2(t) \geq 0$ implies that $S^2 = R^{22} + \mathcal{L}^{2^*}\tilde{Q}^2\mathcal{L}^2$ is positive definite and

$$u^2 = -(S^2)^{-1}\mathcal{L}^{2^*}\tilde{Q}^2(\Phi\tilde{x}_0+\mathcal{L}^1u^1) \tag{4.74}$$

With u^2 given by (4.74), the leader (player 1) has to minimize $\tilde{J}^1(u^1) = J^1(u^1,u^2)$ to determine his optimal control. Again, substituing for $\tilde{x}$ from (4.69) and for u^2 from (4.74), taking gradient of $\tilde{J}^1(u^1)$ w.r.t. u^1 and setting that equal to zero, minimizing u^1 is obtained by

$$S^1u^1 + (L^{1^*}M^{2^*}\tilde{Q}^1M^2\Phi + L^{1^*}\tilde{Q}^2L^2(S^2)^{-1}R^{12}(S^2)^{-1}L^{2^*}\tilde{Q}^2\Phi)\tilde{x}_0 = 0$$

where $S^1 = L^{1^*}M^{2^*}\tilde{Q}^1M^2L^1 + L^{1^*}\tilde{Q}^2L^2(S^2)^{-1}R^{12}(S^2)^{-1}L^{2^*}\tilde{Q}^2L^1 + R^{11}$

$$M^2 = I - L^2(S^2)^{-1}L^{2^*}\tilde{Q}^2 ,$$

I being the identity operator on H. Our assumption that $R^{11} > 0$, $Q_f^1 \geq 0$ and $Q^1(t) \geq 0$ implies that S^1 is positive definite and the optimal controls for the leader and the follower are uniquely given by

$$u^{1^*} = -(S^1)^{-1}L^{1^*}(M^{2^*}\tilde{Q}^1M^2\Phi + \tilde{Q}^2L^2(S^2)^{-1}R^{12}(S^2)^{-1}L^{2^*}\tilde{Q}^2\Phi)\tilde{x}_0 \quad (4.75a)$$

$$u^{2^*} = -(S^2)^{-1}L^{2^*}\tilde{Q}^2(\Phi - L^1(S^1)^{-1}L^{1^*}(M^{2^*}\tilde{Q}^1M^2\Phi + \tilde{Q}^2L^2(S^2)^{-1}R^{12}(S^2)^{-1}L^{2^*}\tilde{Q}^2\Phi))\tilde{x}_0 \quad (4.75b)$$

Clearly, using (4.74) we can write

$$u^{2^*} = -(S^2)^{-1}L^{2^*}\tilde{Q}^2[\tilde{x} - L^2u^{2^*}]$$

implying

$$S^2u^{2^*} - L^{2^*}\tilde{Q}^2L^2u^{2^*} = -L^{2^*}\tilde{Q}^2\tilde{x}$$

and substituting the expression for S^2, we get

$$u^{2^*} = -(R^{22})^{-1}L^{2^*}\tilde{Q}^2\tilde{x} \quad (4.76b)$$

From (4.75a) and using the expression for S^1, we get

$$R^{11}u^{1^*} = -(L^{1^*}M^{2^*}\tilde{Q}^1M^2 + L^{1^*}\tilde{Q}^2L^2(S^2)^{-1}R^{12}(S^2)^{-1}L^{2^*}\tilde{Q}^2)[\tilde{x} - L^2u^{2^*}].$$

Noting that $M^2 = I - L^2(S^2)^{-1}L^{2^*}\tilde{Q}^2$ and using eq^n. (4.76b), we can reassemble all the terms and write

$$u^{1^*} = -(R^{11})^{-1}L^{1^*}(\tilde{Q}^1 - \tilde{Q}^2P)\tilde{x} \quad (4.76a)$$

where the operator P, after some algebraic manipulation, can be shown to satisfy

$$P + L^2(R^{22})^{-1}R^{12}(R^{22})^{-1}L^{2^*}\tilde{Q}^2 - L^2(R^{22})^{-1}L^{2^*}(\tilde{Q}^1 - \tilde{Q}^2P) = 0 \quad (4.76c)$$

Furthermore, substituting (4.76a-b) in (4.69) gives us

$$[I+L^1(R^{11})^{-1}L^{1^*}(\tilde{Q}^1-\tilde{Q}^2P) + L^2(R^{22})^{-1}L^{2^*}\tilde{Q}^2]\tilde{x} = \Phi\tilde{x}_0 \tag{4.76d}$$

Note that the optimal decisions, eq^{n}s. (4.76a-b), are not really closed loop solutions, but merely expressions in closed-loop form of open-loop solutions. In these expressions, $(S^1)^{-1}$ and $(S^2)^{-1}$ do not appear. To obtain more explicit results, we have to calculate the adjoint operators L^{i^*}, i = 1,2. L^i takes $L_2^{m_i}[0,T]$ into H and for an element $w \in H$, we have from definition,

$$\langle w, L^i u^i\rangle = \langle L^{i^*} w, u^i\rangle = \langle u^i, L^{i^*} w\rangle$$

Now

$$\langle w, L^i u^i\rangle = \int_0^T w(t)'\int_0^t \Phi(t,\tau)B^i(\tau)u^i(\tau)d\tau dt$$

$$+ w(T)'\int_0^T \Phi(T,\tau)B^i(\tau)u^i(\tau)d\tau$$

Interchanging the order of integration, we may write

$$\langle w, L^i u^i\rangle = \int_0^T u^i(\tau)'B^i(\tau)'\int_\tau^T \Phi(t,\tau)'w(t)dt \quad d\tau$$

$$+ \int_0^T u^i(\tau)'B^i(\tau)'\Phi(T,\tau)w(T)'d\tau$$

so that $L^{i^*} : H \to L_2^{m_i}[0,T]$ is defined by

$$(L^{i^*} w)(t) = B^i(t)'\int_t^T \Phi(\tau,t)'w(\tau)d\tau + B^t(t)'\Phi(T,t)'w(T), \quad i = 1,2 \tag{4.77}$$

Equations (4.76a-b) can now be expressed as

$$u^{1^*}(t) = -(R^{11})^{-1}B^1(t)'[\int_t^T \Phi(\tau,t)'[Q^1(\tau)-Q^2(\tau)P(\tau)]x(\tau)d\tau$$

$$+ \Phi(T,t)'[Q_f^1-Q_f^2P(T)]x(T) \tag{4.78a}$$

$$u^{2^*}(t) = -(R^{22})^{-1}B^2(t)'[\int_t^T \Phi(\tau,t)'Q^2(\tau)x(\tau)d\tau + \Phi(T,t)'Q_f^2x(T)] \tag{4.78b}$$

Just as in the standard LQ-theory, we can write

$$u^{1^*}(t) = -(R^{11})^{-1}B^1(t)'K^1(t)x(t) \tag{4.79a}$$

$$u^{2^*}(t) = -(R^{22})^{-1}B^2(t)'K^2(t)x(t) \tag{4.79b}$$

where, identifying terms in (4.78a) and (4.79a) and similarly those in (4.78b) and (4.79b), differentiating w.r.t. t, we get

$$\dot{K}^1 = -A'K^1-K^1A-Q^1+Q^2P+K^1B^2(R^{22})^{-1}B^{2'}K^2+K^1B^1(R^{11})^{-1}B^{1'}K^1 \tag{4.80a}$$

$$K^1(T) = Q_f^1 - Q_f^2P(T)$$

$$\dot{K}^2 = -A'K^2-K^2A-Q^2+K^2B^2(R^{22})^{-1}B^{2'}K^2+K^2B^1(R^{11})^{-1}B^{1'}K^1 \tag{4.80b}$$

$$K^2(T) = Q_f^2$$

The matrix-valued function P(t) may be determined from eq^n. (4.76c). In fact, it is given by

$$P(t)x(t) = \int_0^T \Phi(t,\tau)B^2(\tau)(R^{22})^{-1}[B^2(\tau)'K^1(\tau)$$
$$-R^{12}(R^{22})^{-1}B^2(\tau)'K^2(\tau)]x(\tau)d\tau$$

which, on differentiation, gives us

$$\dot{P} = AP-PA+PB^1(R^{11})^{-1}B^{1'}K^1+PB^2(R^{22})^{-1}B^{2'}K^2$$
$$-B^2(R^{22})^{-1}R^{12}(R^{22})^{-1}B^{2'}K^2+B^2(R^{22})^{-1}B^{2'}K^1;\ P(0) = 0 \tag{4.80c}$$

Substituting (4.79a-b) in the state equation gives

$$\dot{x} = (A-B^1(R^{11})^{-1}B^{1'}K^1-B^2(R^{22})^{-1}B^{2'}K^2)x,\ x(0) = x_0 \tag{4.80d}$$

Equations (4.80a-b) are of the Riccati type, but eq^n. (4.80c) is not and it has to be integrated forward in time.

<u>Example 4.3.</u>

We consider a simple model of a dynamic research and development (R&D) funds allocation problem following Chen and Cruz (1972). There are two firms F1 and F2, both competing for the market share of a common item they produce. We assume that market share increases with improved quality of the product which, in turn, is achieved by money spent on R&D. Total market size remains constant during the period under consideration. We want to have an evolution equation describing the technical gap between the two firms. Suppose that firm 2 initially lags behind firm 1. It is clearly easier for firm 2 to catch up with firm 1 by investing more on R&D than for the larger firm 1 to go ahead with more innovation. We model technological gap, denoted by the quantity x, as

$$\dot{x} = \sqrt{r^1} - \alpha\sqrt{r^2} \tag{4.81}$$

where $r^1(t)$ and $r^2(t)$ are the amount of money invested by the two firms on R&D, while $\alpha \geq 1$ indicates the asymmetry of the effect of investment of the two firms as mentioned above. We assume also that when $x = 0$, the two firms have identical share of the market, while if $x = x_0$, the whole market share is taken over by firm 1. Then, over a planning period $[0,T]$, the revenues of the two firms are

$$J^1 = \int_0^T [\tfrac{1}{2}(1+ \frac{x^2}{x_0^2})r - r^1]e^{-\gamma t}dt \tag{4.82a}$$

$$J^2 = \int_0^T [\tfrac{1}{2}(1- \frac{x^2}{x_0^2})r - r^2]e^{-\gamma t}dt \tag{4.82b}$$

where r is a constant and γ is the discount factor. Each firm wants to maximize its revenue J^i and the decision variable available to it is the investment $r^i(t)$ in R&D, $i = 1,2$. We consider only solutions that satisfy $x(t) \leq x_0$. We make the following transformations:

$$y = \bar{e}^{\gamma t/2}x$$

$$u^1 = \sqrt{r^1}\ e^{-\gamma t/2},\ u^2 = \sqrt{r^2}\ e^{-\gamma t/2}$$

$$q = {}^r/2x_0^2$$

Then equations (4.81), (4.82a-b) become

$$\dot{y} = -\frac{\gamma}{2}y + u^1 - \alpha u^2 \tag{4.83}$$

$$J^1 = \int_0^T (qy^2-(u^1)^2)dt + \frac{1}{2}\int_0^T re^{-\gamma t}dt \tag{4.84a}$$

$$J^2 = \int_0^T (-qy^2-(u^2)^2)dt + \frac{1}{2}\int_0^T re^{-\gamma t}dt \tag{4.84b}$$

so that the problem is converted into the LQ-case, with the last term in J^i just constant, $i = 1,2$. We consider the technically lagging firm as the leader, since it has to aggressively compete for its share of the market and the technically advanced firm, which is more secure, will react to this for maintaining its share of the market. Applying the solutions for the optimum strategies as obtained in this section, with firm 2 as the leader, we get

$$u^{1^*} = K^1 y$$

$$u^{2^*} = -\alpha K^2 y \ , \text{ where}$$

$$\dot{K}^1 - \gamma K^1 + (K^1)^2 + q - \alpha^2 K^1 K^2 = 0, \ K^1(T) = 0$$

$$\dot{K}^2 - \gamma K^2 + K^1 K^2 + \alpha^2 (K^2)^2 - q - qP = 0, \ K^2(T) = 0$$

$$\dot{P} + PK^1 + \alpha^2 PK^2 + K^2 = 0, \ P(0) = 0$$

and y(t) satisfies

$$\dot{y} = (-\tfrac{\gamma}{2} + K^1 + \alpha^2 K^2) y$$

Chen and Cruz (1972) made some simulation studies which indicated that the technical gap is rapidly reduced under these open-loop Stackelberg strategies. In fact, the technically advanced firm, in this set-up, does not invest much on R&D with virtually no investment after a while and the technically lagging firm continues investment in R&D for a longer period of time.

5. CONCLUDING REMARKS

To conclude this chapter, we make some comments on feedback Stackelberg solution for 2-person differential games. This problem has been solved, under certain simplifying assumptions, by Papavassilopoulos and Cruz (1979). We do not get into details as it does not contribute directly to the main line of our development. However, some remarks are in order for completeness.

The feedback Stackelberg solution is considerably more involved than the corresponding Nash solution, Consider the same system as described in section 3. Under feedback strategies, equations (4.45)-(4.48) and (4.50) have to be replaced by

$$\dot{\bar{x}} = f(t,\bar{x}(t),u^1(t),\bar{u}^2(t)), \quad \bar{x}(0) = x_0 \tag{4.85}$$

$$\dot{p}_i(t) = -\frac{\partial\phi^2}{\partial x_i}(t,\bar{x},u^1,\bar{u}^2) - \sum_{j=1}^{m_1} \frac{\partial u_j^1(t,\bar{x})}{\partial x_i} \frac{\partial\phi^2(t,\bar{x},u^1,\bar{u}^2)}{\partial u_j^1}$$

$$- \sum_{k=1}^{n} p_k(t)\left(\frac{\partial f_i}{\partial x_k} + \sum_{j=1}^{m_1} \frac{\partial u_j^1}{\partial x_k} \frac{\partial f}{\partial u_j^1}\right)(t,\bar{x},u^1,\bar{u}^2) \tag{4.86}$$

$$p_i(T) = \frac{\partial g^2}{\partial x_i}(\bar{x}(T)) \tag{4.87}$$

$$\frac{\partial\phi^2}{\partial u_j^2}(t,x,u^1,\bar{u}^2) + \sum_{k=1}^{n} p_k(t) \frac{\partial f_k(t,\bar{x},u^1,\bar{u}^2)}{\partial u_j^2} = 0 \quad , \; j = 1,\ldots,m_2 \tag{4.88}$$

The leader has to maximize $W^1(u^1,u^2)$ with respect to u_1 and u_2, under feedback strategies, subject to the constraints (4.85)-(4.88). But this is a completely nonclassical control problem because of the appearence of terms of the form $\frac{\partial u^1}{\partial x_i}$ in the state equations. Under certain assumptions, Papavassilopoulos and Cruz (1979) solved a general non-classical control problem of this type and applied the results obtained to solve the feedback Stackelberg differential game. We do not get into any further details.

CHAPTER 5

REGIONAL INVESTMENT ALLOCATION PROBLEM

1. INTRODUCTION

In this chapter we study the regional investment allocation problem from a different standpoint where conflicts of interest between the regions and between a region and the central planning board are brought explicitly into the model. The choice of criterion for the central planning board is also analyzed in detail. We begin with the model of a dual economy where the poor region has no savings possibilities. We then consider a two-region model where both regions use their savings mechanisms to maximize their respective welfare, resulting in a non-cooperative game problem. Finally, we extend the last model to the general problem of n-regions. The confict between a region and the central planning board results in hierarchical game problems. Mathematical studies of the above mentioned models give rise to a number of interesting questions which are partially answered in this chapter. There still remain a number of outstanding questions which are discussed in detail.

2. INVESTMENT ALLOCATION IN A DUAL ECONOMY

We studied regional investment allocation problem as an application of the maximum principle in chapter 3. We saw that the optimal decisions are "bang-bang" type, with the possibility of at most one switch. This strategy is somewhat academic. In the first place, a complete switch of investment is not implementable in practice. Secondly, the decision to invest all the available resources into either the rich or the poor region is not realistic. We consider in this section a dual economy, consisting of a traditional sector which consumes all it produces and an advanced sector where savings occur which may be invested for economic growth. One obvious objective of the central planning board would be to realise an effective transfer of resources from the advanced to the traditional sector of the economy. The transfer of too much of the resources to the traditional sector is, however, not politically feasible when one realizes that the savings come from the advanced sector only. To incorporate this feature into the model, we consider now the savings incurred by the advanced sector as a control variable for that sector. Thus, we reformulate the problem where the advanced sector has the flexibility to choose its savings strategy once it knows the allocation decision of the central planning board. This problem has been analyzed in Bagchi et al (1981).

We bring labour explicitly into the model. Let us denote the advanced sector of the economy by suffix 1 and the traditional sector by suffix 2. Let K_i denote the capital, L_i the working population and Y_i the output of region i, i = 1,2. Let the production function of region i be denoted by F_i so that

$$Y_i = F_i(K_i, L_i) \tag{5.1}$$

where F_i is assumed to exhibit constant returns to scale with diminishing returns with respect to each factor. Constant returns to scale means $F(\alpha K, \alpha L) = \alpha F(K,L)$, $\alpha > 0$. Diminishing return with respect to each factor means $\partial^2 F/\partial K^2 < 0$ and $\partial^2 F/\partial L^2 < 0$. Suppose that the working population of each region grows at an exogenously determined constant rate n. Different growth rates of population may be easily incorporated into the model. Thus the population L_i grows according as

$$\dot{L}_i = nL_i \ , \ n \text{ constant} \tag{5.2}$$

$$L_i(0) = L_{i0} \text{ given, } i = 1,2$$

Let c_1 denote the fraction of the output of region 1 used for consumption in that region, so that $(1-c_1)F_1(K_1,L_1)$ denote the savings of region 1. Region 2, the traditional sector, consumes all it produces so that $(1-c_1)F_1(K_1,L_1)$ denotes the total savings of the economy. The task of the central planning board is to reallocate this savings between the two regions in a "desirable" manner. Contrary to our analysis in chapter 3, we assume that this allocation decision of the central planning board is constant throughout the planning period. For a short or medium term planning problem, this is a realistic assumption. Let β be the proportion of savings of the region 1 which is reinvested back into the same region, the rest being allocated to the region 2. The balance equations are then given by

$$\dot{K}_1 = \beta F_1(K_1,L_1)(1-c_1), \quad K_1(0) = K_{10} \text{ given} \tag{5.3a}$$

$$\dot{K}_2 = (1-\beta)F_1(K_1,L_1)(1-c_1), \quad K_2(0) = K_{20} \text{ given} \tag{5.3b}$$

We now use the homogeneity of F_1 to write the capital evolution equations in terms of per capita quantities. Denoting K_i/L_i by k_i, $F_i(k_i,1)$ by $f_i(k_i)$ and using eq^n. (5.2), we can write evolution equations for k_i as

$$\dot{k}_1 = \beta f_1(k_1)(1-c_1)-nk_1, \quad k_1(0) = k_{10} \text{ given} \tag{5.4a}$$

$$\dot{k}_2 = (1-\beta)f_1(k_1)(1-c_1)\frac{L_{10}}{L_{20}} - nk_2, \quad k_2(0) = k_{20} \text{ given.} \tag{5.4b}$$

The central planning board chooses β to maximize its objective. But once β is chosen, region 1 has the discretion to choose c_1 so as to maximize its own welfare. The criterion chosen by economists has the general form

$$J_1 = \int_0^T u_1(c_1 f_1(k_1))dt + b_1 k_1(T) \tag{5.5}$$

where T denotes the planning horizon, u_1 is an appropriate utility function and b_1 is a constant. We encountered this type of welfare function while discussing an economic interpretation of the maximum principle in chapter 3. The criterion basically indicates a balance between the total utility of current consumptions and the final capital left at the end of the planning period. The proper choice of the weighting factor b_1 is a difficult one and we discuss one such choice later. Chakravarty (1969) discusses in detail choice of possible welfare criteria in the context of planning problems. The optimal consumption decision of region 1 will naturally depend upon β. Intuitively, it is obvious that as β decreases from 1, region 1 will have less incentive to save, as the savings will then be more and more appropriated to the other region. If β is very low, two situations may arise. If b_1 is dependent on β, the dependence must be such that more and more weight will be given by region 1 to its final capital in the welfare criterion. Thus, with small β, consumption in region 1 will be drastically low throughout the planning period. On the other hand, if b_1 is taken constant and independent of β, the behavioral pattern mentioned above for β decreasing from 1 will persist for small β and region 1 will continue to consume more with decreasing β.

The next question one is faced with is the policy issue confronting the central planning board. This is a difficult question which will be discussed in the next section. We note now that the central planning board has to take basically two factors into account while choosing an appropriate optimizing criterion. One is to reduce economic disparities between the regions at the end of the planning period and the other is to have maximum end capital available for future investment in the economy as a whole. One has to weigh between these two factors while choosing an appropriate criterion for the central planning board. As wil be shown in the next section, one plausible choice of criterion is

$$J = \frac{1}{2}\,\frac{(L_1(T)L_2(T)|k_1(T)-k_2(T)|/(L_1(T)+L_2(T)))+L_{10}k_{10}+L_{20}k_{20}}{L_1(T)k_1(T)+L_2(T)k_2(T)} \qquad (5.6)$$

which the central planning board would like to minimize by an appropriate choice of β.

3. CRITERION OF THE CENTRAL PLANNING BOARD (A DIGRESSION)

Before proceeding further with the resource allocation problem in a dual economy, let us first give a justification for choosing eq^n. (5.6) as a criterion for the central planning board. This is a special case of a more general problem of an economy with n-regions and we discuss the choice of criterion in the context of this general problem. Thus, we have an economy with n regions, region i having capital $K_i(t)$ and population $L_i(t)$, $0 \le t \le T$, $i = 1,\ldots,n$. At the end of the planning period, the central planning board wants to improve capital distribution among the regions (less economic disparity) while, at the same time, making sure that the total end capital for the economy as a whole remains at a reasonable level. We shall try to determine suitable indices for measuring these factors and then, we have to combine them in some manner to arrive at the desired criterion.

We first consider an index for the end capital. Let $k_i(t)$ denote the per-capita capital of region i at time t. Then clearly,

$$K(t) = \sum_{i=1}^{n} L_i(t)k_i(t) \tag{5.7}$$

We choose a simple index for the end capital:

$$\Omega = \frac{K(0)}{K(t)} = \frac{\sum_{i=1}^{n} L_i(0)k_i(0)}{\sum_{i=1}^{n} L_i(T)k_i(T)} \tag{5.8}$$

Maximizing the amount of end capital is equivalent to minimizing Ω. Since capital depreciation has not been taken into the model, $k_i(t)$ is a nondecreasing function of time and the index Ω always lies between 0 and 1.

The next question is the choice of an index that measures inequality of per-capita capital available to the regions at time T. This measure of inequality is important as the central planning board would like to minimize this. There is a vast literature in economics on the question of measuring individual income inequalities. The recent book of Nygård and Sandström (1981) gives a comprehensive account of the present state of knowledge in this vast area of economics. We translate the ideas behind normative measures of individual income inequalities into our context, where we want to measure inequalities of per-capita capital among n regions in an economy.

We begin with some results from the economic literature on measuring individual income inequality. We follow Sen (1976). Consider a community consisting of n persons. All these people are in possession of certain amount of capital and we arrange the people according to their capital endowment:

$y_1 \leq y_2 \leq \ldots \leq y_n$,

where y_i is the capital possessed by the i-th individual. Then the average capital available is

$$\bar{y} = \frac{1}{n} \sum_{i=1}^{n} y_i \tag{5.9}$$

If we want to determine an index that measures inequality in capital possession by members of this community, we would expect that index to satisfy at least the two following axioms:

<u>Axiom 1</u> If all the people are given the same amount of additional capital, the inequality index will decrease.

<u>Axiom 2</u> (Pigou-Dalton condition) If, given the present capital distribution, a transfer of capital is made from a poorer to a richer member of the community, the capital transfer will result in an increase in the inequality index.

Both these axioms are intuitively obvious. Now the Gini-coefficient characterizing the capital distribution of this community and given by

$$G = \frac{1}{2n^2\bar{y}} \sum_{i=1}^{n} \sum_{j=1}^{n} |y_i - y_j| \tag{5.10}$$

satisfies both the axioms stated above. Some elementary algebra enables us to write G in an alternate form:

$$G = \frac{2}{n^2\bar{y}} \sum_{i=1}^{n} (\bar{y} - y_i)(n+1-i) \tag{5.10'}$$

Gini coefficient may be interpreted graphically via the Lorentz curve. Lorentz curve is a graphical illustration of income inequality. It reflects, from the lowest income level upwards, how much percentage of people own what percentage of the total wealth of the community. The Lorentz curve is always convex upwards and the area under the Lorentz curve and its reflection along the line $y = x$ (in the xy-plane) is known to be equal to the Gini coefficient. By way of illustration, we consider a very simple example. Suppose that $y_i = i$, $i = 1,...,n$ so that the ordering of capital possessed by the individuals in our example is automatically satisfied. Then

$$\bar{y} = \frac{n+1}{2}$$

and

$$G = \frac{2}{n^2\bar{y}} \sum_{i=1}^{n} (\bar{y}-y_i)(n+1-i)$$

$$= \frac{2}{n^2\frac{n+1}{2}} \sum_{i=1}^{n} (\frac{n+1}{2} - i)(n+1-i)$$

$$= \frac{1}{3} - \frac{1}{3n}$$

so that for large values of n, $G \cong \frac{1}{3}$.

Let us now turn our attention to the Lorentz curve. The lowest capital possessing group of m people own $\frac{m(m+1)}{2}$ units of capital, while the total population of n people possess $\frac{n(n+1)}{2}$ units of capital. Thus, the lowest capital possessing group forming $\frac{m}{n}$ -th part of the population has in its possession $\frac{m(m+1)}{n(n+1)}$ th part of the total capital. Writing in terms of percentages, we can say, writing $\frac{m}{n}$ to be x, that the lowest $x \times 100\%$ of the capital possessing group has in its possesion $(\frac{n}{n+1} x^2 + \frac{1}{n+1} x) \times 100\%$ of the capital. For n very large, we can say that the lowest $x \times 100\%$ of the capital possessing group has in its possession $x^2 \times 100\%$ of the total capital. This gives us the following Lorentz curve:

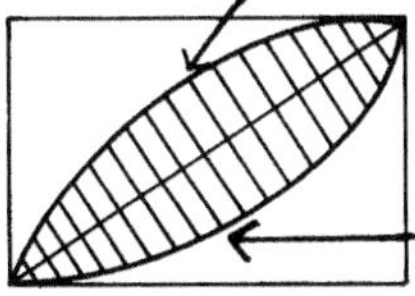

Fig. 5.1 Graphical illustration of Gini coefficient

Now it is obvious that the area between the Lorenz curve and its reflection along the diagonal y = x is given by

$$\int_0^1 (x^{\frac{1}{2}}-x^2)\,dx = \frac{1}{3}$$

which is same as the Gini coefficient for our example.

We now translate this idea of measuring capital or income inequality among individuals to the case of regions. Thus, we consider an economy consisting of n regions, where region i has population L_i, capital K_i and we write $K_i/L_i = k_i$, $i = 1,\ldots,n$. It is the per-capita capital that measures relative welfares of regions. The total working population in our model is

$$L = \sum_{i=1}^{n} L_i \tag{5.11}$$

and per-capita capital for the economy as a whole is given by

$$\bar{k} = \frac{1}{L} \sum_{i=1}^{n} L_i k_i \tag{5.12}$$

In analogy with our previous discussion, we can say that the Gini coefficient for inequality among the n regions is given by

$$G = \frac{1}{2L^2\bar{k}} \sum_{i=1}^{n} \sum_{j=1}^{n} L_i L_j |k_i - k_j| \tag{5.13}$$

What we are basically doing is to consider people in region i as a homogeneous group, each possessing the same capital k_i for $i = 1,\ldots,n$ and then writing the Gini coefficient of capital inequality of all the working people in the economy as a whole. The expression for G given above can be rewritten as

$$G = \frac{1}{L^2\bar{k}} \sum_{i=1}^{n} \sum_{j=1}^{i-1} L_i L_j |k_i - k_j| \tag{5.14}$$

or, equivalently, as

$$G = \frac{\sum_{i=1}^{n} \sum_{j=1}^{i-1} L_i L_j |k_i - k_j|}{\sum_{i=1}^{n} L_i \sum_{i=1}^{n} L_i k_i} \tag{5.14'}$$

As an illustration and for drawing the Lorentz curve, consider an economy consisting of 3 regions, with

$L_1 = 31,\ L_2 = 21,\ L_3 = 1$

and

$k_1 = k,\quad k_2 = 2k,\ k_3 = 3k.$

It is easy to see that, in this case,

$G = \frac{14}{60}$.

Lorentz curve is not smooth in this example and it consists of 3 broken lines. Region 1, the poorest of the region, has 50% of the population and 30% of the total units of capital in its possession. Regions 1 and 2, the two poorest of the regions, together have $\frac{5}{6} \times 100\%$; that is, 83.3% of the population and 70% of the total units of capital in its possession. The following figure illustrates the Lorenz curve for this example:

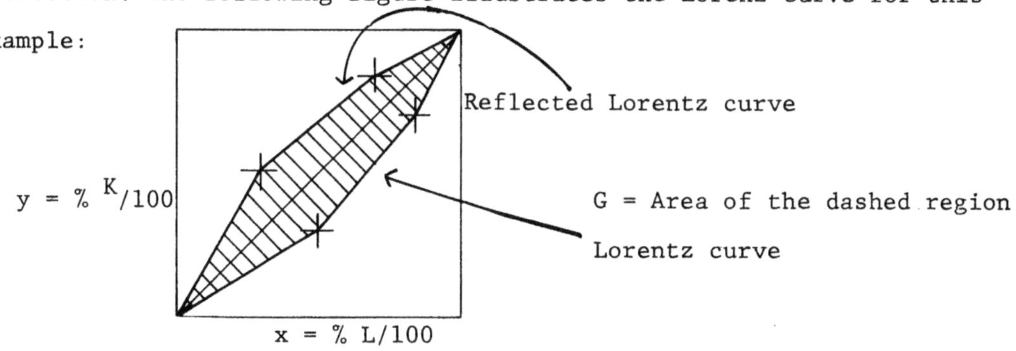

Fig. 5.2 Gini coefficient of per-capita capital inequality.

It is easy to calculate the area of the dashed region which is exactly equal to $\frac{14}{60}$, the Gini coefficient. Examples given in this section are taken from Kuilder (1982).

Now that we have two criteria, one for total capital of the economy as a whole, Ω, and the other for measuring capital inequality among the regions, G, the central planning board has to combine them to arrive at a desired criterion. One obvious choice is the criterion

$$J = \varepsilon G + (1-\varepsilon)\Omega \tag{5.15}$$

where $0 \leq \varepsilon \leq 1$. This criterion lies between 0 and 1 and the objective of the central planning board is to minimize this criterion. More explicitly, we have

$$J = \varepsilon \frac{\sum_{i=1}^{n}\sum_{j=1}^{i-1} L_i(T)L_j(T)\,|k_i(T)-k_j(T)|}{\sum_{i=1}^{n} L_i(T)\sum_{i=1}^{n} L_i(T)k_i(T)} + (1-\varepsilon)\frac{\sum_{i=1}^{n} L_i(0)k_i(0)}{\sum_{i=1}^{n} L_i(T)k_i(T)} \tag{5.16}$$

4. BACK TO THE DUAL ECONOMY

The discussion of the previous section justifies our use of the criterion given by eq^n. (5.6) which is really the special case of eq^n. (5.26) for $n = 2$ and $\varepsilon = 1/2$. For explicitly solving the investment allocation problem in a dual economy, we specify the production functions f_i and the utility function u_1 appearing in the function J_1. Thus, we take Cobb-Douglas production function for both the regions:

$$f_i(k) = A_i k^{\alpha_i} \quad , \quad i = 1,2 \tag{5.17}$$

and take $u_1(x) = \frac{1}{1-\nu}\, x^{1-\nu}$ (5.18)

Let us fix β. We then have an optimization problem at the lower level of the advanced sector (region 1). The state equation is given by

$$\dot{k}_1 = \beta f_1(k_1)(1-c_1) - nk_1 \quad , \quad 0 \le c_1 \le 1 \tag{5.19}$$

$$k_1(0) = K_{10}, \text{ known}$$

Region 1 chooses its own consumption function $c_1(\cdot)$ so as to maximize its own criterion

$$J_1 = \frac{1}{1-\nu}\int_0^T \{c_1 f_1(k_1(t))\}^{1-\nu} dt + b_1 k_1(T) \tag{5.20}$$

We use the maximum principle to obtain a necessary condition for an optimum. The Hamiltonian for the problem is

$$H = \frac{1}{1-\nu}\{c_1 A_1 k_1^{\alpha_1}\}^{1-\nu} + p_1\beta[A_1 k_1^{\alpha_1}(1-c_1)] - p_1 n k_1 \tag{5.21}$$

where the costate p_1 satisfies

$$\dot{p}_1 = -\frac{\partial H}{\partial k_1} = -\alpha_1 (c_1A_1)^{1-\nu} k_1^{\alpha_1(1-\nu)} + p_1\{n - \alpha_1\beta A_1 k_1^{\alpha_1-1}(1-c_1)\}$$

$$p_1(T) = b_1 \tag{5.22}$$

The c_1 dependent part of H is

$$H_1 = \frac{1}{1-\nu}\{c_1A_1k_1^{\alpha}\}^{1-\nu} - p_1\beta A_1k_1^{\alpha_1}c_1.$$

If $p_1 < 0$, H_1 is monotonically increasing as c_1 increases from 0, while if $p_1 > 0$, H_1 has a unique maximum for $c_1 > 0$. Thus, at the optimum, $c_1^* = 0$ in any subinterval is ruled out. The control c_1 is constrained to lie in [0,1] for eacht t. Since H_1 monotonically increases as c_1 increases when $p_1 < 0$, it follows that in this case H_1 will be maximum when $c_1^* = 1$. For $p_1 > 0$, H_1 has a unique maximum for

$$c_1 = (p_1\beta A_1^{\nu} k_1^{\alpha_1\nu})^{-1/\nu} .$$

If the right hand side of the expression above is less than 1, it gives the optimal c_1^*, while in the case that it is greater than or equal to 1, the optimal $c_1^* = 1$ again. Combining all these cases, we see that the optimal

$$c_1^* = \overline{\text{sat}}\,[(p_1\beta A_1^{\nu}k_1^{\alpha_1\nu})^{-1/\nu}], \tag{5.23}$$

where

$$\overline{\text{sat}}\,(x) = x \quad \text{if} \quad 0 \le x < 1$$
$$\qquad\quad = 1 \quad \text{if otherwise.}$$

We have not yet discussed the choice of b_1. It is a complicated question in economic planning in general. One procedure commonly adopted is to determine the steady state (golden rule path) solution of the optimization problem and take b_1 to be close to the price of capital in the steady state. Let us denote the steady state values of k_1 and p_1 by $\bar{k}_1$ and $\bar{p}_1$ respectively. The state and costate equations in the steady state become, with $\bar{c}_1^*$ the corresponding value of c_1^*,

$$\beta A_1\bar{k}_1^{\alpha_1}(1-\bar{c}_1^*) - n\bar{k}_1 = 0$$
$$-\alpha(\bar{c}_1^*A_1)^{1-\nu}\bar{k}_1^{\alpha_1(1-\nu)-1} + \bar{p}_1\{n-\alpha_1\beta A_1\bar{k}_1^{\alpha_1-1}(1-\bar{c}_1^*)\} = 0$$
$$\bar{c}_1^* = \overline{\text{sat}}\,[(\bar{p}_1\beta A_1^{\nu}\bar{k}_1^{\alpha_1\nu})^{-1/\nu}].$$

The solution of this set of equations is

$$\bar{k}_1 = (\alpha_1 \beta A_{1/n})^{1/1-\alpha_1} \quad (5.24a)$$

$$\bar{p}_1 = \beta^{\nu-1}(\beta A_1 \bar{k}_1^{\alpha_1} - n\bar{k}_1)^{-\nu} \quad (5.24b)$$

We, therefore, take $b_1 = \bar{p}_1$.

Note that b_1 in our case depends upon β. Thus, the decision of region 1 regarding the weight b_1 in its criterion functional depends upon the β chosen by the central planning board. For a given β, we have to solve the two-point boundary value problem (5.19) and (5.22) with (5.23) to determine the optimal c_1^* and the corresponding k_1^*. It is also plausible to take b_1 independent of β. Thus, in the real golden rule situation, region2 is expected also to have developed enough so that no transfer of capital from region 1 to region 2 is necessary. Thus, we may assume that $\beta = 1$ in the golden rule situation. In this case, b_1 chosen by region 1 will be independent of β.

To determine the optimum β chosen by the central planning board, we have to know the capital accumulated by the region 2. This is given by the equation

$$\dot{k}_2 = (1-\beta)\frac{L_{10}}{L_{20}} A_1 k_1^{*\alpha_1} (1-c_1^*) - n\, k_2$$

$$k_2(0) = k_{20} \text{ given}$$

Let k_2^* denote the solution of this equation. Then the central planning board has to determine a β that minimizes

$$J = \frac{1}{2}\frac{(L_1(T)L_2(T)|k_1^*(T)-k_2^*(T)|/(L_1(T)+L_2(T)))+L_{10}k_{10}+L_{20}k_{20}}{L_1(T)k_1^*(T)+L_2(T)k_2^*(T)} \quad (5.26)$$

For illustrative purposes, we present some numerical results here. The optimization for β is performed using simple search technique. We use data in a unit comparable to that used by Das (1974) who made a detailed study of an optimal investment planning problem in the framework of the Feldman-Mahalanobis model. Although the units correspond to a planning problem relevant to the Indian economy, the values selected for the initial capitals are somewhat arbitrary:

L_{10} = 10 (number of workers in millions),

L_{20} = 100 (number of workers in millions),

$n = 0.0488$

T = 10 (number of years),

$\nu = 0.6$

$\alpha_1 = 0.6$ $A_1 = 0.32$

$\alpha_2 = 0.4$ $A_2 = 0.149$

k_{10} = 0.1 (in rupees (10,000)

$k_{20} = 0.002$

= 0.007 (in rupees 10,000)

= 0.02

We choose b_1 to be constant, corresponding to the case when $\beta = 1$. The following graph gives the functional relation of J with β corresponding to the three cases: k_{20} = 0.002, 0.007 and 0.02.

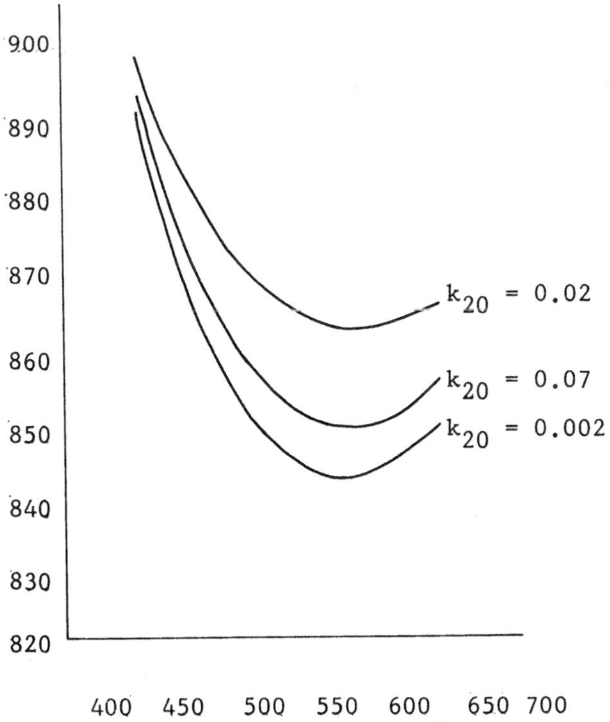

Fig. 5.3 Dependence of J on β.

5. TWO REGION INVESTMENT ALLOCATION PROBLEM

We now consider the general two region investment allocation problem without the restriction that region 2 is too poor to save and consumes all that it produces. Thus, in this model, both regions want to grow optimally according to their own choice, but the savings they make in the process of growth are redistributed by the central planning board with a view to reducing any existing disparity between the regions. The capital evolution equations of the two regions become interconnected via the redistribution mechanism of the central planning board and no region can optimize its own criterion functional independent of the other one. There is an inherent conflict between the two regions in the optimization process and we assume that the regions play a non-cooperative game while trying to maximize their own welfare criteria in the process of growing optimally. At the higher level, the central planning board determines optimal redistribution parameters to maximize the criterion (5.6). The reason for choosing this criterion has been discussed in detail in section 3 of this chapter.

Consider again an economy consisting of two regions which are administratively or geographically separated. Let K_i denote the capital, L_i the working population and Y_i the output of region i, i = 1,2. F_i denotes the production function of region i, so that

$$Y_i = F_i(K_i,L_i), \; i = 1,2 \tag{5.27}$$

We consider the rate of growth of the working population in each region to be exogenously determined and be a constant n. If different regions have different rates of population growth, only minor modification is introduced into the model. Thus

$$\dot{L}_i(t) = n\, L_i(t), \; i = 1,2 \qquad (5.28)$$
$$L_i(0) = L_{i0}$$

Let c_i denote the fraction of the output of region i used for consumtion in that region, so that $(1-c_i)\, F_i(K_i,L_i)$ denote the savings of region i, i = 1,2. The planning board intends to redistribute the savings between the two regions in an appropriate manner to facilitate equitable growth. Suppose that the central planning board allocates a constant fraction $\beta(\gamma)$ of the savings of region 1 (2) back to the same region. The system dynamics is then given by

$$\dot{K}_1 = \beta F_1(K_1,L_1)(1-c_1) + (1-\gamma)F_2(K_2,L_2)(1-c_2) \tag{5.29a}$$

$$K_1(0) = K_{10} \text{ given}$$

$$\dot{K}_2 = (1-\beta)F_1(K_1,L_1)(1-c_1) + \gamma F_2(K_2,L_2)(1-c_2) \tag{5.29b}$$

$$K_2(0) = K_{20} \text{ given}$$

We now use the homogeneity of F_i to write the evolution equation in terms of per capita quantities. We denote per capita quantities by the corresponding small letters and denote $F_i(k_i,1)$ by $f_i(k_i)$. Using $L_r = L_{10}/L_{20}$, we may then write

$$\dot{k}_1 = \beta f_1(k_1)(1-c_1) + (1-\gamma)(1/L_r)f_2(k_2)(1-c_2)-nk_1 \tag{5.30a}$$

$$k_1(0) = k_{10} \text{ given}$$

$$\dot{k}_2 = (1-\beta)L_r f_1(k_1)(1-c_1) + \gamma f_2(k_2)(1-c_2) - nk_2 \tag{5.30b}$$

$$k_2(0) = k_{20} \text{ given.}$$

The central planning board, which acts as a higher level decision maker, chooses β and γ. For fixed β and γ, the local authorities of the two regions act as lower level decision makers and try to optimize their own welfare criteria. Thus, region i wants to choose c_i so as to maximize

$$J_i = \int_0^T u_i(c_i f_i(k_i))dt + b_i k_i(T), \quad i = 1,2 \tag{5.31}$$

where u_i is an appropriate utility function and b_i is a constant, i = 1,2, while T is an appropriate planning horizon. We restrict ourselves to the case when β and γ are constants. For actual planning exercise with short planning horizon, this is a reasonable assumption.

We see now that at the lower level we have two decision makers 1 and

2 who are interested in optimizing their respective criteria J_1 and J_2, subject to the dynamical constraints (5.30a-b). Thus we have a nonzero sum differential game. We assume that the decision makers play non-cooperatively so that it is appropriate to use the Nash solution concept for determining an equilibrium solution for this problem. Some questions of immediate interest arise:

(a) the existence of a Nash equilibrium solution.

(b) the uniqueness of a Nash solution; further narrowing down of the solution concept to guarantee uniqueness.

(c) Existence of a unique maximum of J defined by eq^n. (5.6) for $0 \le \beta, \gamma \le 1$.

Strijbos and Bagchi (1982) first posed and analyzed these questions. We shall discuss these questions and provide partial answers to them in the sequel.

We now solve this problem to determine the consumption patterns of the regions. We take β and γ as prescribed constants. We set $c_2(t)$ a given function and solve the optimization problem for region 1 which will give the optimal $c_1(t)$ as a function of $c_2(t)$. Conversely, we solve the optimization problem for region 2 with a given $c_1(t)$, yielding the optimal $c_2(t)$ as a function of $c_1(t)$. Any solution of this pair of equations gives a Nash solution for the lower (regional) level optimization problem. Let us first fix c_2. The state equations are given by (5.30 a-b) and we have to determine c_1 so that J_1 given by (5.31) is maximized. We use, as before, the maximum principle to obtain the necessary condition for maximum. The Hamiltonian is given by

$$H_1 = u_1(c_1f_1(k_1)) + p_1[\beta f_1(k_1)(1-c_1)+(1-\gamma)(1/L_r)f_2(k_2)(1-c_2)]$$
$$+ q_1[(1-\beta)L_rf_1(k_1)(1-c_1)+\gamma f_2(k_2)(1-c_2)] - n\,(p_1k_1+q_1k_2)$$

where

$$\dot{p}_1 = -u_1'(c_1f_1(k_1))c_1f_1'(k_1)-[\beta p_1+(1-\beta)L_rq_1]f_1'(k_1)(1-c_1)+np_1$$
$$p_1(T) = b_1$$

$$\dot{q}_1 = -[(1-\gamma)(1/L_r)p_1 + \gamma q_1]f_2'(k_2)(1-c_2) + nq_1$$
$$q_1(T) = 0.$$

To determine optimal c_1, we have to solve

$$\frac{\partial H_1}{\partial c_1} = 0.$$

Assuming that u_1' has an inverse, this can be solved to obtain

$$c_1 = u'^{-1}([\beta p_1+(1-\beta)L_r q_1])/f_1(k_1)$$

For c_1 smaller than this value makes $\frac{\partial H_1}{\partial c_1}$ positive. Since the consumption is always bounded between 0 and 1, it readily follows that the optimal

$$c_1 = \min. \{1, u'^{-1}([\beta p_1+(1-\beta)L_r q_1])/f_1(k_1)\}.$$

Analogous results can be obtained for determining optimal c_2 for fixed c_1. We can write down all the equations that must be solved to determine optimal c_1 and c_2 below:

$$\dot{k}_1^* = \beta f_1(k_1^*)(1-c_1^*) + (1-\gamma)(1/L_r)f_2(k_2^*)(1-c_2^*) - nk_1^* \quad (5.32a)$$
$$k_1^*(0) = k_{10}$$

$$\dot{k}_2^* = (1-\beta)L_r f_1(k_1^*)(1-c_1^*) + \gamma f_2(k_2^*)(1-c_2^*) - nk_2^* \quad (5.32b)$$
$$k_2^*(0) = k_{20}$$

$$\dot{p}_1 = -u_1'(c_1^* f_1(k_1^*))c_1^* f_1'(k_1^*) - [\beta p_1+(1-\beta)L_r q_1]f_1'(k_1^*)(1-c_1^*) + np_1 \quad (5.32c)$$

$$p_1(T) = b_1$$

$$\dot{q}_1 = -[(1-\gamma)(1/L_r)p_1 + \gamma q_1]f_2'(k_2^*)(1-c_2^*) + nq_1 \quad (5.32d)$$
$$q_1(T) = 0$$

$$\dot{p}_2 = -u_2'(c_2^* f_2(k_2^*))c_2^* f_2'(k_2^*) - [\gamma p_2+(1-\gamma)(1/L_r)q_2]f_2'(k_2^*)(1-c_2^*) + np_2 \quad (5.32e)$$

$$p_2(T) = b_2$$

$$\dot{q}_2 = -[(1-\beta)L_r p_2+\beta q_2]f_1'(k_1^*)(1-c_1^*) + nq_2 \quad (5.32f)$$
$$q_2(T) = 0$$

$$c_1^* = \min. \{1, u'^{-1}([\beta p_1+(1-\beta)L_r q_1])/f_1(k_1)\} \tag{5.32g}$$

$$c_2^* = \min. \{1, u'^{-1}([\gamma p_2+(1-\gamma)(1/L_r)q_2])/f_2(k_2)\} \tag{5.32h}$$

Equations (5.32 a-h) give the complete set of equations which must be simultaneously solved to determine c_1^* and c_2^*, consumption patterns of the two regions. To get some insight into the nature of the solutions, we consider the simple case of identical regions in the next section.

6. INVESTMENT ALLOCATION FOR TWO IDENTICAL REGIONS

We consider now the situation when both the regions are completely identical. Apparently this is a trivial case, but we can draw some surprising conclusions by studying this simple situation. Thus, we specialize to the case when both the regions have the same initial populations, the same production functions, the same amount of initial capital and identical optimizing criteria. The inherent symmetry of this problem inmmediately implies that the central planning authority has to choose identical reallocation parameters; that is, $\beta = \gamma$. Furthermore, $k_1 = k_2 = k$, $p_1 = p_2 = p$, $q_1 = q_2 = q$, $f_1 = f_2 = f$, $b_1 = b_2 = b$, $c_1 = c_2 = c$, $L_r = 1$ and $u_1 = u_2 = u$. Therefore, the set of equations (5.32 a-h) may now be reduced to the following set of equations:

$$\dot{k} = f(k)(1-c) - nk\colon\ k(0) = k_0 \tag{5.33a}$$

$$\dot{p} = -u'(cf(k)cf'(k) - [\beta p+(1-\beta)q]f'(k)(1-c) + np \tag{5.33b}$$

$$p(T) = b$$

$$\dot{q} = -[(1-\beta)p+\beta q]f'(k)(1-c) + nq;\ q(T) = 0 \tag{5.33c}$$

$$c = \min. \{1, u'^{-1}([\beta p+(1-\beta)q])/f(k)\} \tag{5.33d}$$

Let us first consider the stationary case. For concrete calculations, we specialize to the situation where

$f(x) = Ax^{\alpha}$ (Cobb-Douglas production function)

and

$u(x) = \dfrac{1}{1-\nu}(x-\bar{x})^{1-\nu}$, where $\bar{x}$ is the subsistence level of consumption and $0 < \nu < 1$.

Denoting the steady state value of a variables by an overbar and assuming for the moment that $0 < \bar{c} < 1$, we get

$$0 = A\,\bar{k}^{\alpha}(1-\bar{c}) - n\bar{k}$$

$$0 = -(\bar{c}A\bar{k}^{\alpha}-\bar{x})^{-\nu}\bar{c}A\alpha\bar{k}^{\alpha-1} - \alpha A\bar{k}^{\alpha-1}[\beta\bar{p}+(1-\beta)\bar{q}](1-\bar{c}) + n\bar{p}$$

$$0 = -\alpha[(1-\beta)\bar{p}+\beta\bar{q}]A\bar{k}^{\alpha-1}(1-\bar{c}) + n\bar{q}$$

$$\bar{c} = ([\beta\bar{p}+(1-\beta)\bar{q}]^{-1/\nu}+\bar{x})/A\bar{k}^{\alpha}\ .$$

Solving this set of equations give us

$$\bar{k} = (A(1-\bar{c})/n)^{1/(1-\alpha)} \tag{5.34a}$$

$$\bar{p} = \alpha(A\bar{k}^{\alpha}\bar{c}-\bar{x})^{-\nu} \tag{5.34b}$$

$$\bar{c} = \frac{1-\alpha}{1-\alpha\beta}(1+\alpha-2\alpha\beta) \tag{5.34c}$$

This solution is correct provided we can verify that $\bar{c}$ given by eq^{n}. (5.34c) indeed lies strictly in (0,1). From (5.34c),

$$\bar{c} = 1-\alpha(1+\frac{\alpha(1-\beta)}{1-\alpha\beta}) \geq 1 - \alpha \text{ (since } 0<\alpha<1,\ 0\leq\beta\leq 1)$$

and $\bar{c} = (1-\alpha)\dfrac{1+\alpha-2\alpha\beta}{1-\alpha\beta} \leq (1-\alpha)(1+\alpha) = 1-\alpha^2$ follows from direct verification so that

$$1 - \alpha \leq \bar{c} \leq 1 - \alpha^2$$

implying that $0 < \bar{c} < 1$. Thus $\bar{k}$ given by (5.34a) with $\bar{c}$ given by (5.34c) is indeed the stationary value of the capital in the case of identical regions. It is easy to verify that $\bar{k}$ attains its maximum when $\beta = 1$. Therefore, in the stationary case, the optimal policy for the central planning board is to choose $\beta = \gamma = 1$. In fact, $\bar{k}$ increases monotonically as β increases from 0 to 1. This is a remarkable result, because so long as $\beta = \gamma$, the distribution by the central planning board means that the same amount of resources are transferred from one region to the other and should not apparently make any difference in the planning process. This intuition is contradicted here because of the noncooperative game played at the lower (regional) level. One can argue that even if one region gets back the same amount given from that region to the other by the central planning board, the local (regional) consumption decision is based on an implicit mistrust that the same amount may not be given back to the region and the region reacts to it by consuming more than what it would do otherwise. This reduces the amount that can be transferred to the other region and both the regions become worse off.

Various simulation studies tend to confirm that this feature of the optimal solution remains valid for finite horizon problems as well. We shall later discuss some theoretical analysis that partially confirms this result. We, however, begin with the simulation results. We consider two types of production functions:

(a) Cobb-Douglas Production Function: Suppose that the output is given by

$$Y = F(K,L) = AK^{\alpha}L^{1-\alpha}$$

with

$$A = 0.32, \quad \alpha = 0.6$$

As the criterion of the regional authority, we take

$$J_i = \int_0^T u(c_i f(k_i))dt + b\, k_i(T)$$

where

$$u(x) = \frac{1}{1-\nu}(x-\bar{x})^{1-\nu}$$

with $\nu = 0.6$, b is taken as the price of capital on the golden rule path and

$$n = 0.0488, \; T = 10, \; k_0 = 1, \; \bar{x} = 0.$$

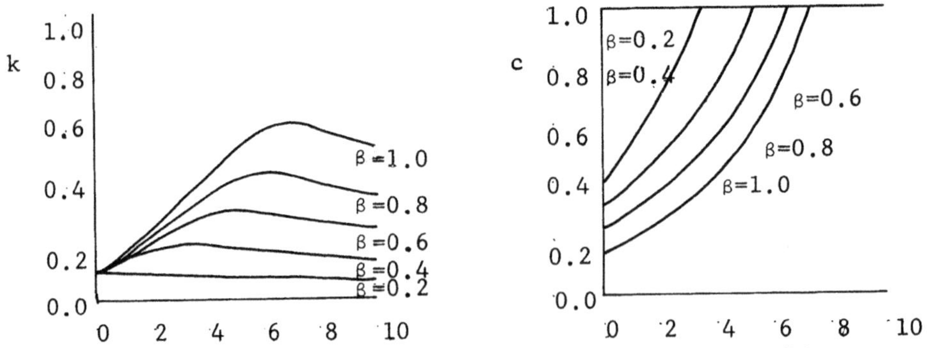

Fig. 5.4(a) Fig. 5.4(b)

Fig. 5.4 Per capita capital (k) and consumption (c) as time functions for different values of β. (Kuilder, 1982).

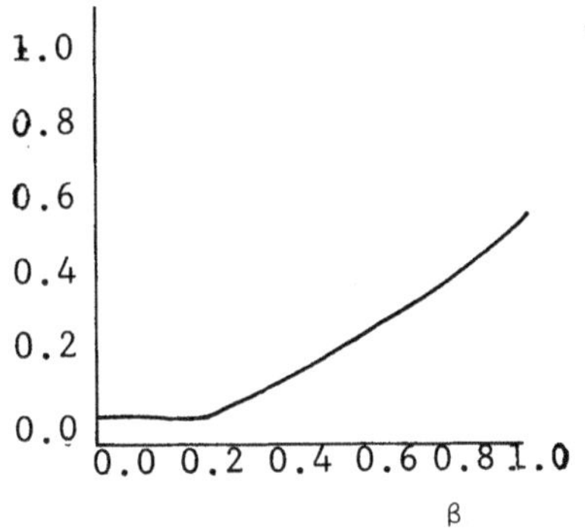

Fig. 5.5(a)

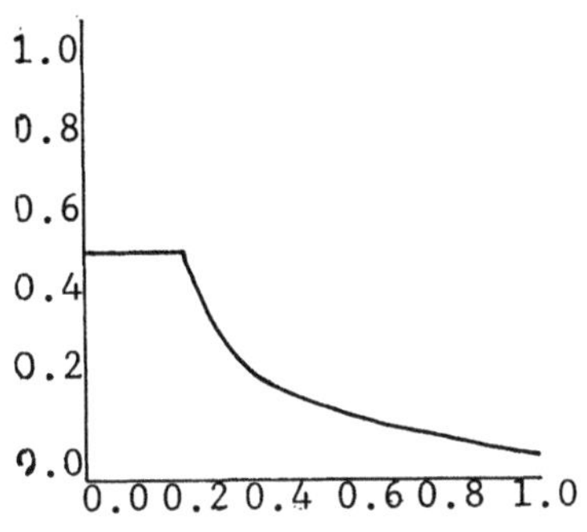

Fig. 5.5(b)

Fig. 5.5 Per capita end capital (k(T)) and the criterion for the central board (J) as functions of β. (Kuilder, 1982).

From the above figures, it is clear that the optimal policy of the central planning board is to give the whole savings of each region back to the same region; that is, to choose $\beta = 1$. Before analyzing this mathematically, let us consider a different kind of production function and see whether this feature of the optimal solution is maintained in that situation as well

(6) C.E.S. Production Function: Suppose that the output is given by

$$Y = F(K,L) = A[BK^{-\alpha}+(1-B)L^{-\alpha}]^{-1/\alpha}$$

with A = 0.32 and B = 0.6. The elasticity of substitution* in this type of production function is $1/(1+\alpha)$ which we denote by σ. With different values of σ, we get the following results:

---

*The elasticity of substitution of a production function is defined by

$$\sigma = \frac{\frac{dL}{dK}/\frac{L}{K}}{d(\frac{dL}{dK})/d(\frac{L}{K})}$$

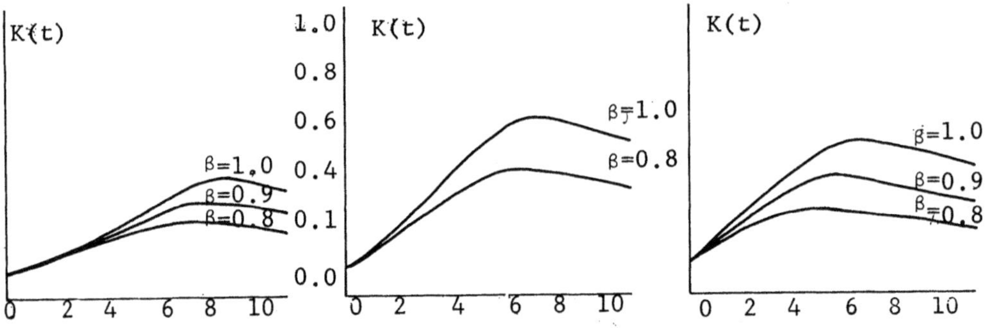

(a) $\sigma = 0.1$ (b) $\sigma = 1.0$ (c) $\sigma = 10.0$

Fig. 5.6 Per capita capital(k) as a function of time voor different values of β and different σ. (Kuilder, 1982).

We see that the general pattern of capital evolution for different values of σ remains similar. In all the different situations, $\beta = 1$ remains the optimal decision rule for the central planning board.

We can infer from the preceding simulation results that $\beta = 1$ is probably the optimal decision rule for the central planning board in the case of an allocation problem for two identical regions. Any redistribution by the central board makes both the regions worse off. But a mathematical proof of this assertion is far from obvious. We give a method of approaching this problem which enables us to obtain at least a partial answer to our basic question of determining the optimal β for the investment allocation problem with two identical regions.
What we are trying to prove is that the end capital k(T) is a monotonically increasing function of β for $\beta \in [0,1]$*. We begin with two lemmas:
Lemma 1. Consider the linear system $\dot{x} = A(t)x$, $t_0 \leq t \leq T$, for a time varying matrix A(t) with continuous components. If all the nondiagonal elements of A(t) consist only of nonnegative elements, the transition matrix $\Phi_A(t,t_0)$ has only nonnegative elements with the diagonal elements being strictly positive for all $t \in [t_0,T]$.

*The main idea behind the following proof is due to R.C.W. Strijbos.

<u>Proof</u> The diagonal elements of A(t) may be negative or zero. The continuity of the elements, however, imply that there exists always an integer $k > 0$ such that

$$\hat{A}(t) = A(t) + kI$$

has only nonnegative elements. It is easy to verify that the transition matrix corresponding to $\hat{A}$ is related to $\Phi_A(t,t_0)$ by the following relation

$$\Phi_A(t,t_0) = \Phi_{\hat{A}}(t,t_0)e^{-kI(t-t_0)} \tag{5.35}$$

Now $\Phi_{\hat{A}}(t,t_0)$ satisfies the following integral equation

$$\Phi_{\hat{A}}(t,t_0) = I + \int_{t_0}^{t} \hat{A}(s)\Phi_{\hat{A}}(s,t_0)ds.$$

This is a Volterra integral equation and if we define the iteration scheme

$$\Phi_{\hat{A}}^{(n)}(t,t_0) = I + \int_{t_0}^{t} \hat{A}(s)\Phi_{\hat{A}}^{(n-1)}(s,t_0)ds$$

$$\Phi_{\hat{A}}^{(0)}(t,t_0) = I \tag{5.36}$$

we know that it always converges to the true solution; that is,

$$\Phi_{\hat{A}}^{(n)}(t,t_0) \longrightarrow \Phi_{\hat{A}}(t,t_0) \text{ as } n \to \infty \; \forall t \in [t_0,T].$$

Since $\hat{A}(s)$ consists only of nonnegative elements for all s, eq^n. (5.36) readily implies by induction that all the nondiagonal elements of $\Phi_{\hat{A}}^{(n)}(t,t_0)$ are nonnegative and all the diagonal elements are ≥ 1. Since $\Phi_{\hat{A}}^{(n)}(t,t_0) \to \Phi_{\hat{A}}(t,t_0)$, as $n \to \infty$, the same conclusion holds for the elements of $\Phi_{\hat{A}}(t,t_0)$ as well. Eq^n, (5.35) then implies that all the nondiagonal elements of $\Phi_A(t,t_0)$ are nonnegative and all the diagonal elements are strictly positive. □

<u>Lemma 2</u>

Consider the linear differential equation

$$\dot{x} = A(t)x + b(t), \quad 0 \leq t \leq T$$

where all the nondiagonal elements of A(t) are nonpositive and all the components of b(t) are nonnegative. Suppose that x(t) has dimension n and

$$x_1(0) \geq 0$$

$$x_i(T) = 0 \ , \ i = 2,\ldots,n.$$

Then

$$x_1(T) \geq 0.$$

<u>Proof</u> Suppose that $x_1(T) < 0$. Define

$$y(t) = x(T-t).$$

Then

$$\dot{y}(t) = (-A)y(t) + (-b)$$

with

$$y(0) = x(T) = \begin{bmatrix} -ve \\ 0 \\ \vdots \\ 0 \end{bmatrix};$$

-i.e., the first component of y(0) is strictly negative while the other components are zero. Then

$$y(T) = \Phi_{-A}(T,0)y(0) + \int_0^T \Phi_{-A}(T-s)(-b)ds.$$

From our assumption and using Lemma 1 all the components of the integral on the right hand side above are nonpositive. Since the 11-th element of $\Phi_{-A}(T,0)$ is strictly positive and y(0) has its first component strictly negative while the other components are zero, it follows that

$$y_1(T) < 0$$

That is,

$$x_1(0) < 0.$$

This contradicts our assumption that $x_1(0) \geq 0$ implying that $x_1(T) \geq 0$.

□

Our objective is to prove that k(T) increases monotonically with respect to β for $\beta \in [0,1]$. Thus, we try to prove that $k_\beta(T) = \frac{\partial k(T)}{\partial \beta} \geq 0$ for $\beta \in [0,1]$. We have to work with equations (5.33 a-d) to arrive at our desired conclusion. Here the production function f and the utility function u are very general. We make the following assumptions:

$$f(x) > 0, \quad f'(x) > 0, \quad f''(x) < 0 \quad \text{for } x > 0 \tag{5.37a}$$

$$u(x) > 0, \quad u'(x) > 0, \quad u''(x) < 0 \quad \text{for } x > 0 \tag{5.37b}$$

and obviously $b > 0$.

Lemma 3

In equations (5.33 a-d), with $k_0 \geq 0$, $k(t) \geq 0$, $p(t) \geq 0$, $q(t) \geq 0$ and $p(t) - q(t) \geq 0$ for $0 \leq t \geq T$.

Proof That $k(t) \geq 0$ for all t follows readily from eq^n. (5.33a). To prove the other inequalities, define

$$x = (p \quad q \quad p-q)'.$$

Then

$$\dot{x} = A(t)x + b(t)$$

where

$$A = \begin{bmatrix} n-\beta f'(k)(1-c) & -(1-\beta)f'(k)(1-c) & 0 \\ -(1-\beta)f'(k)(1-c) & n-\beta f'(k)(1-c) & 0 \\ 0 & 0 & n-(2\beta-1)f'(k)(1-c) \end{bmatrix}$$

$$b = (-u'(cf(k))cf'(k) \qquad 0 \qquad -u'(cf(k))cf'(k))'$$

and $x(T) = (b \quad 0 \quad 0)'$.

Define

$$y(t) = x(T-t)$$

Then

$$\dot{y}(t) = (-A)y + (-b)$$

All the nondiagonal elements of $(-A)$ are nonnegative and all the components of $(-b)$ are nonnegative. Then

$$y(t) = \Phi_{-A}(t,0)y(0) + \int_0^t \Phi_{-A}(t,s)(-b)ds.$$

By lemma 1, all the components of $\Phi_{-A}(t,s)$ are nonnegative and $y(0) = x(T)$ has only nonnegative components.

Therefore, all the componenets of $y(t)$ are also nonnegative establishing the lemma. □

Let us now use eq^{n}s. (5.33 a-d) to write down equations for k_β, p_β and q_β. For this, we treat the two cases $c = 1$ and $c \neq 1$ separately.

If c = 1,

$$\dot{k}_\beta = -nk_\beta \tag{5.38a}$$

$$\dot{p}_\beta = np_\beta - [u''(f(k))\ f'(k)^2 + u'(f(k))f''(k)]k_\beta \tag{5.38b}$$

$$\dot{q}_\beta = nq_\beta \tag{5.38c}$$

while if $c \neq 1$, we may rewrite (5.33 a-d) as

$$\dot{k} = f(k) - u'^{-1}([\beta p+(1-\beta)q]) - nk \tag{5.39a}$$

$$\dot{p} = -[\beta p+(1-\beta)q]f'(k) + np \tag{5.39b}$$

$$\dot{q} = -[(1-\beta)p+\beta q][f'(k) - \frac{f'(k)}{f(k)} u'^{-1}([\beta p+(1-\beta)q])]+ nq \tag{5.39c}$$

If we define

$$\psi = \beta p + (1-\beta)q$$

and $\phi = (1-\beta)p + \beta q$,

we can write equations (5.39 a-c) equivalently as

$$\dot{k} = f(k) - u'^{-1}(\psi) - nk \tag{5.40a}$$

$$\dot{p} = -\psi f'(k) + np \tag{5.40b}$$

$$\dot{q} = -\phi[f'(k) - \frac{f'(k)}{f(k)} u'^{-1}(\psi)] + nq \tag{5.40c}$$

We can now write down differential equations for k_β, p_β and q_β :

$$\dot{k}_\beta = [f'(k)-n]k_\beta - u'^{-1'}(\psi)\psi_\beta \tag{5.41a}$$

$$\dot{p}_\beta = np_\beta - \psi f''(k)k_\beta - f'(k)\psi_\beta \tag{5.41b}$$

$$\begin{aligned}\dot{q}_\beta = {} & nq_\beta - \phi_\beta f'(1-c) + \phi\frac{f'(k)}{f(k)} u'^{-1'}(\psi)\psi_\beta \\ & -\phi\ [f''(k) - \frac{f'(k)^2}{f(k)}(1-c) + \frac{f'(k)^2}{f(k)}]k_\beta\end{aligned} \tag{5.41c}$$

$$\psi_\beta = p - q + \beta p_\beta + (1-\beta)q_\beta \tag{5.41d}$$

$$\phi_\beta = -p + q + (1-\beta)p_\beta + \beta q_\beta \tag{5.41e}$$

Since $k(0) = k_0$, $p(T) = b$ and $q(T) = 0$, follows that

$$k_\beta(0) = 0 \tag{5.42a}$$

$$p_\beta(T) = 0 \tag{5.42b}$$

$$q_\beta(T) = 0 \tag{5.42c}$$

Finally, we note that $u'^{-1}(x) > 0$ and $u'^{-1'}(x) < 0$ for $x > 0$.

Theorem 5.1

k(T) has a local maximum at $\beta = 1$.

Proof We have to show that $k_\beta(T)|_{\beta=1} > 0$. When $\beta = 1$, we get

$$\psi = p,\ \phi = q,\ \psi_\beta = p + p_\beta,\ \phi_\beta = -p + q_\beta$$

since

$$q(t)|_{\beta=1} = 0.$$

Then equations (5.41 a-c) become

$$\dot{k}_\beta = [f'(k)-n]k_\beta - u'^{-1'}(p)(p+p_\beta) \qquad (5.43a)$$

$$\dot{p}_\beta = np_\beta - pf''(k)k_\beta - f'(k)(p+p_\beta) \qquad (5.43b)$$

Define

$$x = (k_\beta - p_\beta)'.$$

Then

$$\dot{x} = A(t)x + b(t)$$

where, for c = 1, from equations (5.38 a-b)

$$A = \begin{bmatrix} -n & 0 \\ u''(f(k))f'(k)^2+u'(f(k))f''(k) & n \end{bmatrix}, \quad b = \begin{bmatrix} 0 \\ 0 \end{bmatrix}$$

while, for $c \neq 1$, from equations (5.43 a-b)

$$A = \begin{bmatrix} f'(k)-n & u'^{-1'}(p) \\ pf''(k) & n-f'(k) \end{bmatrix}, \quad b = \begin{bmatrix} -pu'^{-1'}(p) \\ pf'(k) \end{bmatrix}$$

We see that in both cases, nondiagonal elements of A are nonpositive and components of b are nonnegative.

Furthermore, $k_\beta(0) = 0$ and $p_\beta(T) = 0$.

From lemma 2, we conclude that

$$k_\beta(t) \geq 0 \quad \text{for } \beta = 1$$

What we really want to show, of course, is that k(T) has a global maximum for $\beta = 1$; that is, $k_\beta(T) \geq 0$ for all β in [0,1]. A sufficient condition for this to hold is given by the following:

Theorem 5.2

Suppose that

$$f''(k) + \frac{f'(k)^2}{f(k)} \leq 0.$$

Then $k_\beta(T) \geq 0$ for $\beta \in [1/2, \ 1]$.

<u>Proof</u> Let us now define

$$x = (k_\beta - p_\beta \ -(p_\beta + q_\beta))'.$$

Then we can write

$$\dot{x} = A(t)x + b(t)$$

where A(t) and b(t) have very complicated expressions.

For c = 1,

$$A = \begin{bmatrix} -n & 0 & 0 \\ u''(f(k))f'(k)^2 + u'(f(k))f''(k) & n & 0 \\ u''(f(k))f'(k)^2 + u'(f(k))f''(k) & 0 & n \end{bmatrix}, \quad b = \begin{bmatrix} 0 \\ 0 \\ 0 \end{bmatrix}$$

while for $c \neq 1$,

$$A = \begin{bmatrix} f'(k)-n & (2\beta-1)u'^{-1'}(\psi) & (1-\beta)u'^{-1'}(\psi) \\ \psi f''(k) & n-(2\beta-1)f'(k) & -(1-\beta)f'(k) \\ \psi f''(k) + \phi[f''(k)(1-c)+c\frac{f'(k)^2}{f(k)}] & -(2\beta-1)cf'(k) + \phi\frac{f'(k)}{f(k)}u'^{-1'}(\psi)(2\beta-1) & n-f'(k)[(1-\beta)+\beta(1-c)] + \phi\frac{f'(k)}{f(k)}u'^{-1'}(\psi)(1-\beta) \end{bmatrix}$$

and

$$b = \begin{bmatrix} -(p-q)u'^{-1'}(\psi) \\ f'(k)(p-q) \\ f'(k)c(p-q) - \phi\frac{f'(k)}{f(k)}u'^{-1'}(\psi)(p-q) \end{bmatrix}$$

For all c, we see that b(t) has only nonnegative components. For c = 1, the nondiagonal elements of A are nonpositive. The situation is, however, complicated when $c \neq 1$. In this case, all the nondiagonal elements of A are clearly nonpositive except the 31-th element, provided $\beta \geq 1/2$. The 31-th element is given by

$$\begin{aligned} a_{31} &= \psi f''(k) + \phi[f''(k)(1-c)+c\frac{f'(k)^2}{f(k)}] \\ &= \psi f''(k) + \phi[(f''(k) - \frac{f'(k)^2}{f(k)})(1-c) + \frac{f'(k)^2}{f(k)}] \\ &= \phi(f''(k) - \frac{f'(k)^2}{f(k)})(1-c)+(\psi-\phi)f''(k) + \phi(f''(k)+\frac{f'(k)^2}{f(k)}) \end{aligned}$$

$$= \phi(f''(k) - \frac{f'(k)^2}{f(k)})(1-c) + (2\beta-1)(p-q)f''(k)$$

$$+ \phi(f''(k) + \frac{f'(k)^2}{f(k)}).$$

If $f''(k) + \frac{f'(k)^2}{f(k)} \leq 0$, a_{31} is also nonpositive and we can apply lemma 2. Thus under this sufficient condition,

$k_\beta(T) \geq 0$ for $\beta \geq 1/2$ □

This is only a partial solution to our original assertion. The approach taken here, although interesting, is somewhat ad hoc in the sense that the choice of an appropriate vector x so that lemma 2 may be applied is really a guesswork. It is interesting to determine a consistent procedure for attacking such problems. This is similar to the problem of sensitivity analysis in optimal control problems. In the economics literature, Brock (1971) and Oniki(1973) considered similar problems and unfortunately, their methods of attack were also not general, but were rather problem dependent.

7. TWO REGION INVESTMENT ALLOCATION PROBLEM (CONTINUED)

Let us now turn to the general problem of nonidentical regions. Our analysis of the previous section confirms that any unnecessary redistribution of capital is disadvantageous for both the regions. This leads us to conjecture that in the general case when one optimizes J over the square $0 \leq \beta \leq 1$, $0 \leq \gamma \leq 1$, the optimum β and γ will always lie on the boundaries of the square $\beta = 1$ or $\gamma = 1$. A mathematical analysis of this assertion is considerably more difficult than in the case of identical regions for two reasons. Firstly, one has to now work with two independent parameters instead of one and secondly, the criterion J in this case is considerably more complicated (in the case of identical regions, $|k_1-k_2| = 0$ always). This is a challenging problem in hierarchical control and we state precisely the conjecture below:

Problem At the lower level, a noncooperative differential game is played where the state dynamics is given by equations (5.30 a-b) and player i wants to maximize J_i given by eq^n. (5.31) for i = 1,2. At the higher level, the decision maker wants to choose β and γ, $0 \le \beta, \gamma \le 1$, so that

$$J(\beta,\gamma) = \frac{1}{2} \frac{(L_1(T)L_2(T)|k_1(T)-k_2(T)|/(L_1(T)+L_2(T)))+L_{10}k_{10}+L_{20}k_{20}}{L_1(T)k_1(T)+L_2(T)k_2(T)}$$

is minimized. The conjecture states that J attains its minimum at a point on the boundary $\beta = 1$ or $\gamma = 1$.

Let us note an interesting consequence in case our conjecture is indeed correct. If the optimum really lies on the boundary $\beta = 1$ or $\gamma = 1$, the game problem at the lower level reduces to a sequence of two control problems. If, for example, $\gamma = 1$, region 1 can optimize its consumption pattern independent of region 2. The optimal c_1 and k_1 thus obtained can then be substituted in the optimization problem for region 2 where they appear only as exogenously determined quantities. The existence and uniqueness of Nash equilibrium solution is then assured using standard results in control theory.

Extensive simulation studies have been undertaken to ascertain the validity of this conjecture. The table below gives one typical simulation result.

β \ γ	1.00	0.98	0.96	0.94	0.92	0.90
1.00	0.281	0.279	0.278	0.276	0.275	0.274
0.99	0.284	0.282	0.280	0.279	0.278	0.278
0.98	0.286	0.284	0.283	0.282	0.281	0.281

TABLE 5.1 Dependence of J on β and γ

In the simulation, we consider the same Cobb-Douglas production function for both the regions:

$$Y = F(K,L) = AK^{\alpha}L^{1-\alpha}$$

with

$$A = 0.32, \alpha = 0.6$$

The utility function $u(\cdot)$ is taken to be the same for both the regions and having the form:

$$u(x) = \frac{1}{1-\nu} x^{1-\nu}$$

with $\nu = 0.6$.

b is taken to be the price of capital on the golden rule path. Furthermore, the following data are used:

$$T = 10, \quad n = 0.0488, \quad L_r = 0.5, \quad k_{10} = 0.1, \; k_{20} = 1.$$

8. n-REGION INVESTMENT ALLOCATION PROBLEM

We can readily generalize the previous analysis to the case of n regions. As before, the production of region i is denoted by

$$Y_i = F_i(K_i,L_i), \; i = 1,\ldots,n \tag{5.44}$$

with K_i the capital and L_i the population of region i. The population grows exogenously as

$$\dot{L}_i = nL_i, \; L_i(0) = L_{i0} \text{ given} \tag{5.45}$$

With c_i denoting the fraction of the production used for consumption by region i, the savings of the region is given by

$$S_i = F_i(K_i,L_i)(1-c_i) \tag{5.46}$$

Now the central planning board receives the savings of each region and redistributes it to all the regions. Suppose that out of the savings of region i, a fraction $\gamma_{i,j}$ is given to region j for investment in that region. The amount of investment capital received from the central planning board is then clearly

$$I_i = \sum_{j=1}^{n} \gamma_{j,i} S_j, \qquad \sum_{i=1}^{n} \gamma_{j,i} = 1 \tag{5.47}$$

Ignoring capital depreciation, we see that the capital evolution equation of region i is given by

$$\dot{K}_i = I_i = \sum_{j=1}^{n} \gamma_{j,i} F_j(K_j, L_j)(1-c_j) \qquad (5.48)$$

with $K_i(0) = K_{i0}$ given $i = 1, \ldots, n$.

An before, writing $k_i = K_i/L_i$ and assuming homogeneity of the production function, we can write

$$F_i(K_i, L_i)/L_i = F_i(k_i, 1) = f_i(k_i), \text{ say.}$$

We can then write down the evolution equation for k_i as

$$\dot{k}_i = \sum_{j=1}^{n} \{\gamma_{j,i} \frac{L_j}{L_i} f_j(k_j)(1-c_j)\} - nk_i. \qquad (5.49)$$

$$k_i(0) = k_{i0} \text{ given.}$$

For given parameters $\gamma_{i,j}$, each region wants to maximize its own welfare:

$$J_i = \int_0^T u_i(c_i f_i(k_i))dt + b_i \, k_i(T) \qquad (5.50)$$

similar to the case of two regions. Among the regions, therefore, a game is played. Assuming that the game is played in a noncooperative manner, we try to determine a Nash equilibrium and call it a solution of the lower (regional) level optimization problem. The optimum c_i^*'s naturally depend on the parameters $\gamma_{i,j}$, $i,j = 1, \ldots, n$ $(\sum_{j=1}^{n} \gamma_{i,j} = 1)$. The problem of the central planning board is to determine the parameters $\gamma_{i,j}$ so that

$$J = \varepsilon \frac{\sum_{i=1}^{n} \sum_{j=1}^{i-1} L_i(T) L_j(T) |k_i(T) - k_j(T)|}{\sum_{i=1}^{n} L_i(T) \sum_{i=1}^{n} L_i(T) k_i(T)} + (1-\varepsilon) \frac{\sum_{i=1}^{n} L_i(0) k_i(0)}{\sum_{i=1}^{n} L_i(T) k_i(T)} \qquad (5.51)$$

is minimized

Using the maximum principle, we obtain the following set of equations which the optimum c_i's and k_i's must satisfy:

$$\dot{k}_i = \sum_{j=1}^{n} \gamma_{j,i} \frac{L_j}{L_i} f_j(k_j) - nk_i \; ; \; k_i(0) = k_{i0} \tag{5.52a}$$

$$\dot{p}_{i,i} = -u_i' \; (c_i f_i(k_i)) c_i f_i'(k_i) - (\sum_{m=1}^{n} \gamma_{i,m} \frac{L_i}{L_m} p_{i,m}) f_i'(k_i)(1-c_i) + np_{i,i} \tag{5.52b}$$

$$\dot{p}_{i.j} = - (\sum_{m=1}^{n} \gamma_{j,m} \frac{L_j}{L_m} p_{i,m}) f_j'(k_j) + np_{i,j} \tag{5.52c}$$

$$p_{i,i}(T) = b_i \; ; \; p_{i,j}(T) = 0, \; i,j = 1,\ldots,n \text{ with } i \neq j \tag{5.52d}$$

$$c_i = \min . \{1, u'^{-1}_i (\sum_{m=1}^{n} \gamma_{i,m} \frac{L_i}{L_m} p_{i,m})/f_i(k_i)\} \tag{5.52e}$$

We can readily extend our conjecture of the previous section to the case of n-regions. What the conjecture really asserts is that it is always suboptimal to allow any "unnecessary" exchange of resources between two regions. Extending this to the case of n-regions, we may conjecture that for the optimal decision of the central planning board, the following rule must previal: "any region which has to give up a part of its savings during the redistribution process cannot get any parts of the savings of other regions". In terms of $\gamma_{i,j}$'s, we can formulate this as follows:

For the optimal decision of the central planning board,

$$\gamma_{i,i} \neq 1 \Rightarrow \gamma_{j,i} = 0 \text{ for } j = 1,\ldots,n, \; j \neq i.$$

We, of course, have always the additional constraint

$$\sum_{j=1}^{n} \gamma_{i,j} = 1, \quad i = 1,\ldots,n.$$

This expresses mathematically our assertion that the central planning board should not indulge in any unnecessary capital transfers among the regions. We can also equivalently say that the optimum of J is attained only on some edges of the parallelopiped $0 \leq \gamma_{i,j} \leq 1$, $j \neq 1$, $i,j = 1,\ldots,n$. A typical edge on which the optimum will lie is characterized by $\gamma_{1,j} = 0$, $j = 2,\ldots,n$; $\gamma_{2,j} = 0$, $j = 3,\ldots,n$;...; $\gamma_{n-1,n} = 0$.

If this conjecture holds, it readily follows that at the lower (regional) level, for γ's on such edges, the game problem reduces to a sequence of n optimal control problems. The existence and uniqueness of the Nash equilibrium solution will be assured for those γ's.
The following simulation study with 3- regions agree with our assertion. We consider the same production function and welfare criterion for the regions:

$$f_i(k_i) = Ak_i^{\alpha}, \quad A = 0.32, \quad \alpha = 0.60$$

$$J_i = \int_0^T u(c_i f_i(k_i))dt + b_i k_i(T)$$

with

$$u(x) = \frac{1}{1-\nu}(x-\bar{x})^{1-\nu}, \quad \nu = 0.60.$$

We take $b_i = 1.00065$, $n = 0.0488$, $T = 10.0$, $\bar{x} = 0$.
The central planning board wants to minimize J given by (5.51) with $\varepsilon = 1/2$ and $n = 3$. The first term in J is the Gini coefficient which we denote by G and the second term gives a measure of end capital which we denote by Ω. We consider the following initial conditions:

$$k_1(0) = 0.1, \quad k_2(0) = 0.2, \quad k_3(0) = 0.4$$

$$L_1(0) = L_2(0) = L_3(0) = 1.0.$$

The optimal $\gamma_{i,j}$'s are given by the following matrix:

$$\Gamma = \begin{bmatrix} \gamma_{1,1} & \gamma_{1,2} & \gamma_{1,3} \\ \gamma_{2,1} & \gamma_{2,2} & \gamma_{2,3} \\ \gamma_{3,1} & \gamma_{3,2} & \gamma_{3,3} \end{bmatrix} = \begin{bmatrix} 1.00 & 0.00 & 0.00 \\ 0.03 & 0.97 & 0.00 \\ 0.14 & 0.00 & 0.86 \end{bmatrix}$$

This tallies fully with our assertion that

$$\gamma_{i,i} = 1 \Rightarrow \gamma_{i,j} = 0, \; j = 1,2,3, \quad j \neq i$$

and

$$\sum_{j=1}^{3} \gamma_{i,j} = 1, \quad i = 1,2,3.$$

We can also study the variation in G and Ω separately. The following table gives the values of the Gini-coefficient in the neighbourhood of the optimum:

$\gamma_{1,1}$	$\gamma_{1,2}$	$\gamma_{1,3}$	$\gamma_{2,1}$	$\gamma_{2,2}$	$\gamma_{2,3}$	$\gamma_{3,1}$	$\gamma_{3,2}$	$\gamma_{3,3}$	G
1.00	0.00	0.00	0.03	0.97	0.00	0.14	0.00	0.86	0.001
1.00	0.00	0.00	0.03	0.97	0.00	0.13	0.00	0.87	0.006
1.00	0.00	0.00	0.03	0.97	0.00	0.15	0.00	0.85	0.007
1.00	0.00	0.00	0.03	0.97	0.00	0.13	0.01	0.86	0.002
1.00	0.00	0.00	0.03	0.97	0.00	0.14	0.01	0.85	0.007
1.00	0.00	0.00	0.03	0.97	0.00	0.15	0.01	0.84	0.011
1.00	0.00	0.00	0.04	0.97	0.00	0.14	0.00	0.86	0.008
1.00	0.00	0.00	0.02	0.98	0.00	0.14	0.00	0.86	0.006
1.00	0.00	0.00	0.03	0.96	0.01	0.14	0.00	0.86	0.007
0.99	0.0	0.01	0.03	0.97	0.00	0.14	0.00	0.86	0.006
0.99	0.01	0.00	0.03	0.97	0.00	0.14	0.00	0.86	0.006

TABLE 5.2 Dependence of Ω on $\gamma_{i,j}$; i,j = 1,2,3

We saw in the case of two regions that J decreases monotically with both β and γ. This feature of the solution remains in this example with 3-regions as well. Thus, in our example, Ω decreases monotonically with $\gamma_{1,1}$, $\gamma_{2,2}$ and $\gamma_{3,3}$.

Before concluding this section, we give a graph theoretic characterization of the optimal decisions, γ's, of the central planning board. Consider an economy consisting of 3 regions with $k_1(0) > k_2(0) > k_3(0)$. We denote a region by a node of a graph and the capital flow (via redistribution) from one region to another by a directed line. There are three possibilities by which the central planning board may redistribute the savings among the regions. Firstly, it is possible that for the optimal J, the richest region has to share parts of its savings with both the two remaining regions. Secondly, the richest region may have to share its savings only with the poorest one, with the region in the middle growing independently. Finally, the richest and the next better-off region may both have to share their savings with the poorest region. We illustrate this by means of the following diagram:

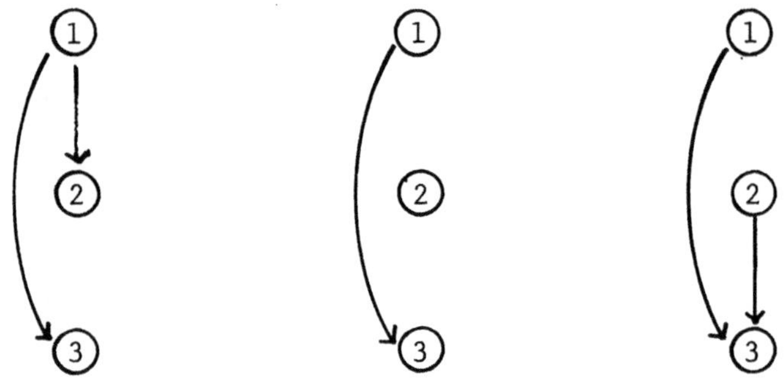

Fig. 5.7 Possible optimal redistribution rules for the case of 3 regions

Directed lines in any other direction are not possible in the optimal decision state of the central planning board. The same feature remains in the case of n-regions. It is not possible for two nodes to be connected by directed lines in both the directions.

Most of the material presented in sections 6-8 may be found in Kuilder (1982). It also contains a study of some alternate redistribution mechanisms. The model may be made more realistic by allowing for migration and introducing tax mechanism into it. The conjecture stated in section 7 is a challenging problem in hierarchical control theory.

CHAPTER 6

STACKELBERG GAMES IN LINEAR PROGRAMMING CONTEXT

1. INTRODUCTION

So far we have been mainly concerned with the situations where the interactions between the decisions are modelled by means of differential or difference equations. In economic applications, however, problems are often modelled in a linear programming setting. In this chapter, we study these types of Stackelberg games. We formulate the problem precisely in the next section. We then study, in detail, properties of the reaction curve in this context. Finally, we use the properties of the reaction curve to develop an algorithm for solving the Stackelberg LP-problem.

2. STACKELBERG LINEAR PROGRAMMING (SLP) PROBLEM

We restrict our attention only to the case of two players. A SLP problem for two players is defined if the following three conditions hold:

(C1) The two players (the leader and the follower) select values for their decision variables (u_1 and u_2, respectively) in such a way that both minimize their respective linear objective functions $J_1(u_1,u_2) = c_{11}'u_1 + c_{12}'u_2$ and $J_2(u_1,u_2) = c_{21}'u_1 + c_{22}'u_2$.

(C2) The decisions of the players are interconnected by means of a set of inequalities $A_1u_1 + A_2u_2 \le b$ together with a set of nonnegativity constraints $u_1 \ge 0$, $u_2 \ge 0$.

(C3) The leader announces his decision and the follower is compelled to take this into account while minimizing his objective function $J_2(u_1,u_2)$ subject to the constraints.

Note that, in general, $u_i \in \mathbb{R}^{m_i}$, i=1,2, so that if b is a m-vector, A_i are $m \times m_i$-matrices, i=1,2,. The inequality signs stand for inequalities componentwise.

As expected, we are faced with a two stage optimization problem. For a fixed decision u_1^o of the leader, the follower is faced with the following problem:

$$\min c_{21}'u_1^o + c_{22}'u_2 \qquad \text{(P1)}$$

subject to the constraint

$$A_2u_2 \leq b - A_1u_1^o, \; u_2 \geq 0.$$

Under certain conditions, the follower's problem has a unique solution for every admissible u_1 (i.e., all possible u_1 for which there exists at least one u_2 s.t. $A_1u_1 + A_2u_2 \leq b$). We denote this unique solution of the follower by $u_2^*(u_1)$ corresponding to a decision u_1 of the leader. Then the leader is faced with the following problem:

$$\min. c_{11}'u_1 + c_{12}'u_2 \qquad \text{(P2)}$$

subject to the constraint

$$A_1u_1 + A_2u_2 \leq b, \; u_1 \geq 0, \; u_2 \geq 0,$$

$$u_2 = u_2^*(u_1).$$

The decision of the leader has m_1 components and that of the follower has m_2 components. Let us now define some sets which are useful in studying the SLP problem.

<u>Definition 6.1</u>

The subset FS of $\mathbb{R}^{m_1+m_2}$ defined by

$$FS = \{(u_1,u_2)\in\mathbb{R}^{m_1+m_2} \mid A_1u_1+A_2u_2\leq b, u_1\geq 0, u_2\geq 0\} \qquad (6.1)$$

is called the feasible set for the problem, where $A_1 \in \mathbb{R}^{m\times m_1}$, $A_2 \in \mathbb{R}^{m\times m_2}$ and $b \in \mathbb{R}^m$. We assume that this set FS is nonempty, as otherwise our SLP problem would be meaningless. FS is convex polytope in $\mathbb{R}^{m_1+m_2}$.

<u>Definition 6.2.</u>

The subset R of $\mathbb{R}^{m_1}$ defined by

$$R = \{u_1\in\mathbb{R}^{m_1} \mid \exists u_2\in\mathbb{R}^{m_2} \text{ s.t. } (u_1,u_2)\in FS\} \qquad (6.2)$$

is called the <u>range</u> of the leader. The leader has to announce a decision belonging to this set R. Otherwise, the follower will be unable to react to it and choose his own decision u_2 such that (u_1,u_2) becomes feasible. The assumption that FS is nonempty implies that R is also a nonempty set. Clearly, R is the orthogonal projection of FS on $\mathbb{R}^{m_1}$ and is, therefore, again a convex polytope in $\mathbb{R}^{m_1}$.

<u>Definition 6.3</u>

For a given $u_1 \in R$, the subset $FRS(u_1)$ of $\mathbb{R}^{m_2}$ defined by

$$FRS(u_1) = \{u_2 \in \mathbb{R}^{m_2} \mid A_2u_2 \le b - A_1u_1, u_2 \ge 0\} \tag{6.3}$$

is called the <u>feasible reaction set</u> of the follower corresponding to a decision u_1 of the leader. It follows from the previous discussion that $FRS(u_1)$ is also nonempty for any $u_1 \in R$.

<u>Definition 6.4</u>

For a given $u_1 \in R$, the subset $ORS(u_1)$ of $FRS(u_1)$ is defined by

$$ORS(u_1) = \{u_2^* \in FRS(u_1) \mid J_2(u_1,u_2^*) = \min_{u_2 \in FRS(u_1)} J_2(u_1,u_2)\} \tag{6.4}$$

is called the optimal reaction set of the follower corresponding to a decision u_1 of the leader. Although the follower is free to choose any element from $FRS(u_1)$, he will naturally choose an element that minimizes $J_2(u_1,u_2)$; i.e., he will choose an element from $ORS(u_1)$. So, $ORS(u_1)$ is the set of solutions of the linear programming problem formulated in (P1) above, also referred to as the follower's problem.

In formulating the leader's problem (P2) above, what we have really assumed is that $ORS(u_1)$ is a singleton. The minimization problem occurring in the definition 6.4 is a feasible LP-problem so long as $FRS(u_1)$ is nonempty. Our assumption, therefore, means that the LP-problem occurring in the definition 6.4 has a unique bounded solution. Let us, for the time being, relax this assumption. In general, the solution may not be unique and may not even be bounded in which case $ORS(u_1)$ is an empty set. In this latter situation, there must exist a $d \in \mathbb{R}^{m_2}$ for which

$A_2d \leq 0$, $d \geq 0$ and $c_{22}'d < 0$ hold. But this would imply that the follower's problem has an unbounded solution or, equivalently, that ORS(u_1) is an empty set for every $u_1 \in R$. Thus, in general, a dichotomy prevails: the optimal reaction set of the follower is either nonempty for all $u_1 \in R$ or is empty for all $u_1 \in R$. The latter situation is not interesting and we rule that out. Thus we make the following

Assumption 1

The feasible set FS is not empty. The optimal reaction sets of the follower are not empty; that is, for all $d \in \mathbb{R}^{m_2}$ with $A_2d \leq 0$ and $d \geq 0$, $c_{22}'d \geq 0$ must hold.

Now the leader knows that if he announces his u_1 the reaction of the follower will be selection of an element in ORS(u_1). The pair of points determined by $u_1 \in R$ and $u_2 \in$ ORS(u_1) plays a central role in the Stackelberg problem.

Definition 6.5

The reaction curve for our problem is defined by

$$RC = \{(u_1,u_2) \in \mathbb{R}^{m_1+m_2} \mid u_1 \in R, u_2 \in ORS(u_1)\} \subset FS \tag{6.5}$$

Because of its crucial role in the Stackelberg problem, we study the properties of RC in detail.

3. PROPERTIES OF THE REACTION CURVE

In the special case when $c_{22} = 0$ (that is, the follower is not interested in his own decision problem), ORS(u_1) = FRS(u_1) for all $u_1 \in R$ and this implies that RC = FS. We, therefore, assume that $c_{22} \neq 0$, since otherwise RC is known completely. We denote the i-th row of A_j by A_{ji} in the following discussion.

Definition 6.6

By a face F of the convex polytope FS we mean a part of the boundary of FS. More specifically,

$$F = \{(u_1,u_2)|u_{1i}=0(i=i_1,\dots,i_p),\ u_{2i} = 0(i=i_{p+1},\dots,i_q),$$

$$A_{1i}u_1 + A_{2i}u_2 = b_i(i=i_{q+1},\dots,i_r)\} \cap FS$$

with $0 \le p \le q \le r$ and at least one of these three inequalities holding with strict inequality is a face of FS. If this intersection is not empty, the dimension of the face F is, by definition, equal to $m_1+m_2-\text{rank}.A^F$, where A^F is the matrix consisting of

$e_{i_1},\dots,e_{i_p},e_{i_{p+1}+m_1},$

$\dots,e_{i_q+m_1},\ [A_{1i_{q+1}},A_{2i_{q+1}}],\dots,[A_{1i_r},A_{2i_r}]$, where

e_i stand for the i-th unit vector in $\mathbb{R}^{m_1+m_2}$.

We now study the properties of the reaction curve RC.

Property 1

Assume that $c_{22} \neq 0$. If $(u_1^0,u_2^0) \in RC$, (u_1^0,u_2^0) must lie on a face of FS; that is, at least one of the constraints $u_{2i} \ge 0$, $i=1,\dots,m_2$ and $A_{1i}u_1 + A_{2i}u_2 \le b_i$, $i=1,\dots,m$ is satisfied with an equality.

Proof. Suppose, on the contrary, that $(u_1^0,u_2^0) \in RC$ and that $u_2^0 > 0$ and $A_1u_1^0 + A_2u_2^0 < b$. It is easy to see that there is an ε-neighbourhood of u_2^0, every element u_2 of which satisfies $u_2 > 0$ and

$$A_2u_2 < b - A_1u_1^0.$$

Thus, any u_2 with $||u_2-u_2^0|| < \varepsilon$ belongs to $FRS(u_1^0)$. Since $(u_1^0,u_2^0) \in RC$, it follows that $J_2(u_1^0,u_2^0) \le J_2(u_1^0,u_2)$ for u_2 in the ε-neighbourhood of u_2^0. Since $c_{22} \neq 0$, there is at least one nonzero component of c_{22}, say c_{22k}. Let us assume, without loss of generality, that $c_{22k} > 0$. With e_k the k-th unit vector in $\mathbb{R}^{m_2}$, we then have

$$\begin{aligned} J_2(u_1^0,u_2^0-\tfrac{1}{2}e_k) &= c_{21}'u_1^0 + c_{22}'u_2^0 - \tfrac{1}{2}\varepsilon c_{22}'e_k \\ &= J_2(u_1^0,u_2^0) - \tfrac{1}{2}\varepsilon c_{22k} < J_2(u_1^0,u_2^0) \end{aligned}$$

contradicting our assumption that $(u_1^0,u_2^0) \in RC$. This proves our assertion. □

Definition 6.7

Define index sets $I^0(u_1,u_2)$, $I^1(u_1,u_2)$ and $I^2(u_1,u_2)$ for any feasible point (u_1,u_2) by

$$i \in I^0(u_1,u_2) \Longleftrightarrow A_{1i}u_1 + A_{2i}u_2 = b_i$$

$$i \in I^1(u_1,u_2) \Longleftrightarrow u_{1i} = 0$$

$$i \in I^2(u_1,u_2) \Longleftrightarrow u_{2i} = 0$$

Property 1 may be rephrased as follows: If $c_{22} \neq 0$ and $(u_1^0,u_2^0) \in RC$, at least one of the index sets $I^0(u_1^0,u_2^0)$ and $I^2(u_1^0,u_2^0)$ is not empty.

Property 2

Assume $c_{22} \neq 0$. $(u_1^0,u_2^0) \in RC$ if and only if any feasible point (u_1,u_2) satisfying $I^j(u_1^0,u_2^0) \subset I^j(u_1,u_2)$, $j = 0,2$, belongs to RC.

Proof

Suppose that $(u_1^0,u_2^0) \in RC$ and $(u_1^+,u_2^+) \in FS$ satisfying $I^0(u_1^0,u_2^0) \subset I^0(u_1^+,u_2^+)$ and $I^2(u_1^0,u_2^0) \subset I^2(u_1^+,u_2^+)$, but $(u_1^+,u_2^+) \notin RC$. Then there exists a $u_2^* \in FRS(u_1^+)$ such that $c_{22}'u_2^* < c_{22}'u_2^+$.
$u_2^* \in FRS(u_1^+) \Longrightarrow A_2u_2^* \le b - A_2u_1^+$. The inclusion relations above then imply that

$$\forall i \in I^0(u_1^0,u_2^0),\ A_{2i}(u_2^+-u_2^*) = b_i - A_{1i}u_1^+ - A_{2i}u_2^* \ge 0 \tag{6.6a}$$

$$\forall i \in I^2(u_1^0,u_2^0),\ u_{2i}^+ - u_{2i}^* = -u_{2i}^* \le 0 \tag{6.6b}$$

and we have just concluded that

$$c_{22}'(u_2^+-u_2^*) \ge 0 \tag{6.6c}$$

Just as before, it is easy to see that there is a ε-neighbourhood $N_\varepsilon(u_2^0)$ such that

(a) $\forall u_2 \in N_\varepsilon(u_2^0)$ and $\forall i \notin I^0(u_1^0,u_2^0)$,

$$A_{2i}u_2 < b_i - A_{1i}u_1^0 \tag{6.7a}$$

(b) $\forall u_2 \in N_\varepsilon(u_2^0)$ and $\forall i \notin I^2(u_1^0,u_2^0)$,

$$u_{2i} > 0 \tag{6.7b}$$

Take any τ such that $0 < \tau < \varepsilon/\|u_2^+ - u_2^*\|$. Then $u_2^0 - \tau(u_2^+-u_2^*) \in N_\varepsilon(u_2^0)$ and equations (6.6a-b), (6.7a-b) imply that

(a) $\forall i \in I^0(u_1^0,u_2^0)$, $A_{2i}(u_2^0-\tau(u_2^+-u_2^*)) \le A_{2i}u_2^0 = b_i - A_{1i}u_1^0$

(b) $\forall i \notin I^0(u_1^0,u_2^0)$, $A_{2i}(u_2^0-\tau(u_2^+-u_2^*)) < b_i - A_{1i}u_1^0$

(c) $\forall i \in I^2(u_1^0,u_2^0)$, $u_{2i}^0 - \tau(u_{2i}^+-u_{2i}^*) \ge u_{2i}^0 = 0$

(d) $\forall i \notin I^2(u_1^0,u_2^0)$, $u_{2i}^0 - \tau(u_{2i}^+-u_{2i}^*) > 0$.

Therefore, $u_2^0 - \tau(u_2^+-u_2^*) \in FRS(u_1^0)$. But using (6.6c) we see that

$$J_2(u_1^0,u_2^0-\tau(u_2^+-u_2^*))=c_{21}'u_1^0+c_{22}'u_2^0-\tau c_{22}'(u_2^+-u_2^*)<J_2(u_1^0,u_2^0)$$

contradicting the assumption that $(u_1^0,u_2^0) \in RC$. Thus (u_1^+,u_2^+) must belong to RC. The converse is trivially established by taking $(u_1,u_2) = (u_1^0,u_2^0)$. □

<u>Property 3</u>

Assume $c_{22} \neq 0$. The reaction curve RC is the union of a finite number of faces F_i of the convex polytope FS; that is,

$$RC = \bigcup_{i\in I} F_i, \quad I_F = \{i|F_i \subset RC\} \tag{6.8}$$

<u>Proof</u> Since FS is determined by a finite number of inequalities the result is an immediate consequence of properties 1 and 2. □

<u>Remark</u> We cannot directly use the characterization of RC given in property 3. For large problems, it is impractical to generate all the

faces F_i in (6.8). However, the result gives an insight into the nature of the reaction curve. Although every face of the feasible set is convex, the reaction curve is not, in general, convex.

Property 4

Assume $c_{22} \neq 0$. Convex combinations of feasible points not in RC do not belong to RC. Strict convex combinations of a feasible point not in RC and a point in RC do not belong to RC.

Proof Suppose that $(u_1,u_2) \in FS$, $(v_1,v_2) \in FS$ but $(u_1,u_2) \notin RC$. Consider a convex combination

$$(w_1,w_2) = \lambda(u_1,u_2) + (1-\lambda)(v_1,v_2), \quad 0 < \lambda < 1. \tag{6.9}$$

One can easily verify that

$$I^j(w_1,w_2) = I^j(u_1,u_2) \cap I^j(v_1,v_2), \ j = 0,2.$$

Suppose that $(w_1,w_2) \in RC$. Since $I^j(w_1,w_2) \subset I^j(u_1,u_2)$, $j = 0,2$ and $(u_1,u_2) \in FS$, property 2 implies that $(u_1,u_2) \in RC$, contradicting our assumption. Thus, $(w_1,w_2) \notin RC$.

Suppose, in addition, that $(v_1,v_2) \notin RC$. Then $(w_1,w_2) \notin RC$ for $0 \leq \lambda \leq 1$, where (w_1,w_2) is as defined by (6.9). □

Property 5

If $(u_1,u_2) \in RC$, $(v_1,v_2) \in RC$ and $I^j(u_1,u_2) \subset I^j(v_1,v_2)$ or $I^j(v_1,v_2) \subset I^j(u_1,u_2)$ for $j = 0,2$, any convex combination of (u_1,u_2) and (v_1,v_2) belongs to RC.

Proof Follows readily from property 2 and the proof of property 4. □

Remark

We cannot, in general, assert more than what property 5 says about the reaction curve. Even adjacent extreme points of FS, both lying in RC, may have their convex combination not being in RC.

The following simple three dimensional example illustrates the properties of the reaction curve.

Example 6.1*

Consider a three-dimensional SLP problem described by the figure below. Player P1, the leader, chooses $x_1 \in \mathbb{R}$ while the player P2, the follower, chooses $(x_2,x_3) \in \mathbb{R}^2$ such that (x_1,x_2,x_3) belongs to the convex polyhedron, the feasible set FS, spanned by A, B, C and D.

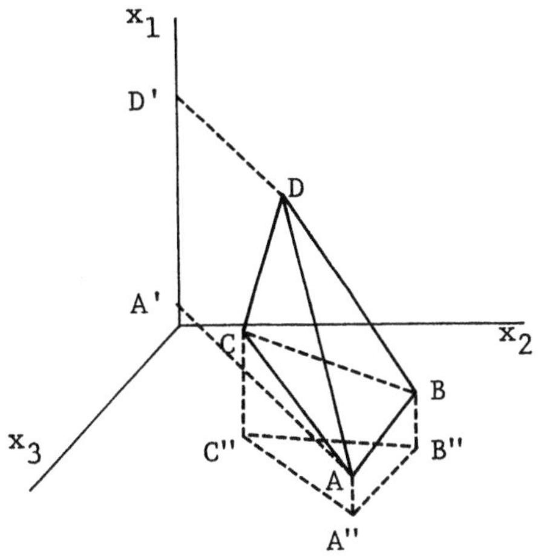

Fig. 6.1 Illustration of feasible sets and reaction curves.

The range of P1 is the orthogonal projection of ABCD on the x_1-axis, which in this case is the closed interval A'D'. FRS(u) for any u $\in$ A'D' is the intersection of the hyperplane $x_1 = u$ with FS. With proper choice of d_2 and d_3 in the criterion $J_2(x_1,x_2,x_3) = d_1x_1 + d_2x_2 + d_3x_3$, ORS(u) for u $\in$ A'D' will be singletons on the line segments AB, BC and CD. Thus, the reaction curve is the union of three one-dimensional edges (faces) of the tetrahedron ABCD; in fact, RC = AB $\cup$ BC $\cup$ CD.

Properties 1, 2, 4 and 5 stated above of the reaction curve may be easily verified in this example. Although A, B and C are adjacent

*This example is due to M. Moraal.

extreme points of FS, their convex combinations may or may not be elements of RC. The problem changes drastically if we change the roles of the players P1 and P2. If P2 is the leader, its range is the triangle A"B"C", the orthogonal projection of ABCD on the (x_2,x_3)-plane. FRS(v) for any $v \in$ A"B"C" is then a closed interval. If $c_1 > 0$ in $J_1(x_1,x_2,x_3) = c_1x_1 + c_2x_2 + c_3x_3$, the reaction curve consists of only one face of FS; namely, ABC. If $c_1 < 0$, the reaction curve is the union of three faces of FS; namely, ABD $\cup$ ACD $\cup$ BCD. The properties 1, 2, 4 and 5 may be easily verified in this case as well.

So far, we have allowed the possibility that ORS(u_1) contains more than one element. We now make the following

Assumption 2

The optimal reaction set of the follower is a singleton for all $u_1 \in R$; or, equivalently, that the follower's problem has a unique optimal solution for all $u_1 \in R$.

Remark

Under assumptions 1 and 2, the optimization problem for the leader may be rephrased as

$$\min.\ c_{11}'u_1 + c_{12}'u_2 \qquad \text{(P2')}$$

such that $(u_1,u_2) \in RC$.

It is clear that the Stackelberg linear programming problem is a minimization problem of a linear criterion over a nonconvex set. We study some features of this minimization problem below.

Definition 6.8

A team solution for the leader is a minimum point of the leader's criterion on the feasible set; or, more specifically, any solution to the problem

$$c_{11}'u_1 + c_{12}'u_2 \qquad \text{(P3)}$$

subject to $(u_1,u_2) \in FS$

is called a team solution for the leader.

Remark

In the degenerate case when $c_{22} = 0$, we have seen that RC = FS and so, problems (P2') and (P3) are equivalent. Thus, in this trivial case, Stackelberg solutions and Team solutions for the leader are equivalent.

Theorem 6.1

If a Team solution for the leader exists, a Stackelberg solution also exists. Moreover, if there is a team solution (u_1^*, u_2^*) for the leader on the reaction curve, it is also a Stackelberg solution and any Stackelberg solution is then a Team solution for the leader.

Proof Because of the remark above, we may take $c_{22} \neq 0$ and assume that a Team solution for the leader exists so that J_1 is bounded from below on FS. Minimization of J_1 on each of the (nonempty) faces of FS is then a feasible and bounded LP-problem. There are only a finite number of faces of all possible dimensions. We take one optimal solution of each of these problems and arrange them so that the corresponding value of J_1 is arranged in a nondecreasing order. The first optimal solution in this ordered set that lies on the reaction curve is a Stackelberg solution for the problem. Suppose there is no optimal solution in this arrangement that lies on the reaction curve. If optimal solution on a face does not lie on RC, the whole face must be disjoint from RC (property 3) implying that RC = $\emptyset$, contradicting our assumptions. This proves that there is always a Stackelberg solution for the problem. Suppose now that a Team solution for the leader, (u_1^*, u_2^*), lies on the reaction curve. The existence of a Team solution implies the existence of a Stackelberg solution which we denote by (u_1^τ, u_2^τ). Then

$$(u_1^*, u_2^*) \in RC, \; J_1(u_1^*, u_2^*) \leq J_1(u_1, u_1) \quad \forall (u_1, u_2) \in FS$$

$$(u_1^\tau, u_2^\tau) \in RC \subset FS, \; J_1(u_1^\tau, u_2^\tau) \leq J_1(u_1, u_2) \quad \forall (u_1, u_2) \in RC.$$

It follows that

$$J_1(u_1^*,u_2^*) = J_1(u_1^\tau,u_2^\tau) \le J_1(u_1,u_2) \quad \forall\,(u_1,u_2) \in RC$$

and

$$J_1(u_1^\tau,u_2^\tau) = J_1(u_1^*,u_2^*) \le J_1(u_1,u_2) \quad \forall\,(u_1,u_2) \in FS.$$

It follows that (u_1^*,u_2^*) is itself a Stackelberg solution and (u_1^τ,u_2^τ) is a Team solution for the leader by definition. □

Theorem 6.2

If a Stackelberg solution exists, there is always a Stackelberg solution at an extreme point of the feasible set.

Proof If $c_{22} = 0$, the result holds because Stackelberg solutions are then precisely Team solutions and the Team problem is a standard LP-problem which has always solution at an extreme point. We consider therefore $c_{22} \ne 0$. Suppose that Stackelberg solution (u_1^τ,u_2^τ) exists. Since FS has extreme points, it can be written as

$$FS = \{(u_1,u_2)\,|\,(u_1,u_2) = \sum_{i=1}^{P} \alpha_i (u_1^{ei},u_2^{ei}) + \sum_{j=1}^{Q} \beta_j (v_1^{rj},v_2^{rj}) \tag{6.10}$$

$$\text{with} \quad \sum_{i=1}^{P} \alpha_i = 1 \text{ and } \alpha_i,\beta_j \ge 0,\ i=1,\ldots,P,\ j=1,\ldots,Q,$$

where $\{(u_1^{ei},u_2^{ei})\,|\,i=1,\ldots,P\}$ is the set of extreme points of FS and $\{(v_1^{rj},v_2^{rj})\,|\,j=1,\ldots,Q\}$ is the set of extreme rays of FS (that is, (v_1^{rj},v_2^{rj}) satisfies $A_1v_1^{rj} + A_2v_2^{rj} \le 0$, $v_1^{rj} \ge 0$, $v_2^{rj} \ge 0$, $j=1,\ldots,Q$).

Since $(u_1^\tau,u_2^\tau) \in FS$, there exists $\alpha_i^\tau,\beta_j^\tau \ge 0$, $\sum_{i=1}^{P} \alpha_i^\tau = 1$ such that

$$(u_1^\tau,u_2^\tau) = \sum_{i=1}^{P} \alpha_i^\tau (u_1^{ei},u_2^{ei}) + \sum_{j=1}^{Q} \beta_j^\tau (v_1^{rj},v_2^{rj}) \tag{6.12}$$

Since $\sum_{i=1}^{P} \alpha_i^\tau = 1$, there is always one index k with $\alpha_k^\tau > 0$. Assume that $0 \lessgtr \alpha_k^\tau \lessgtr 1$. Define

$$(u_1^{pk},u_1^{pk}) = \frac{1}{1-\alpha_k^\tau} \{ \sum_{\substack{i=1\\ i\ne k}}^{P} \alpha_i^\tau (u_1^{ei},u_2^{ei}) + \sum_{j=1}^{Q} \beta_j^\tau (v_1^{rj},v_2^{rj}) \}$$

$$= \sum_{\substack{i=1\\ i\neq k}}^{P} \frac{\alpha_i^{\tau}}{1-\alpha_k^{\tau}} (u_1^{ei}, u_2^{ei}) + \sum_{j=1}^{Q} \frac{\beta_j^{\tau}}{1-\alpha_k^{\tau}} (v_1^{rj}, v_2^{rj}) \tag{6.12}$$

Clearly

$$\frac{\alpha_i^{\tau}}{1-\alpha_k^{\tau}} \geq 0 \ (i\neq k,\ i=1,\dots,P); \quad \frac{\beta_j^{\tau}}{1-\alpha_k^{\tau}} \geq 0 \ (j=1,\dots,Q)$$

and

$$\sum_{\substack{i=1\\ i\neq k}}^{P} \frac{\alpha_i^{\tau}}{1-\alpha_k^{\tau}} = 1.$$

Therefore, $(u_1^{pk}, u_2^{pk}) \in FS$ (follows from (6.10) and (6.12)). Now we can write

$$(u_1^{\tau}, u_2^{\tau}) = \alpha_k^{\tau}(u_1^{ek}, u_2^{ek}) + (1-\alpha_k^{\tau})(u_1^{pk}, u_2^{pk}),\ 0 < \alpha_k^{\tau} < 1$$

where (u_1^{ek}, u_2^{ek}) is also in FS. Thus, (u_1^{τ}, u_2^{τ}) is a strict combination of two feasible points. Since $(u_1, u_2) \in RC$, it follows from property 4 that (u_1^{ek}, u_2^{ek}) and (u_1^{pk}, u_2^{pk}) must both belong to RC. Furthermore,

$$J_1(u_1^{\tau}, u_2^{\tau}) = \alpha_k^{\tau}\ J_1(u_1^{ek}, u_2^{ek}) + (1-\alpha_k^{\tau}) J_1(u_1^{pk}, u_2^{pk})$$

$$\geq \alpha_k^{\tau}\ J_1(u_1^{\tau}, u_2^{\tau}) + (1-\alpha_k^{\tau}) J_1(u_1^{\tau}, u_2^{\tau}) = J_1(u_1^{\tau}, u_2^{\tau}) \tag{6.13}$$

since (u_1^{τ}, u_2^{τ}) is a Stackelberg solution and $(u_1^{ek}, u_2^{ek}) \in RC$ and $(u_1^{pk}, u_2^{pk}) \in RC$. It follows from (6.13) that $J_1(u_1^{ek}, u_2^{ek}) = J_1(u_1^{\tau}, u_2^{\tau})$, $(u_1^{ek}, u_2^{ek}) \in RC$ and is an extreme point of FS, proving our assertion. If there is no index k such that $0 < \alpha_k^{\tau} < 1$, there is an index k with $\alpha_k^{\tau} = 1$ and $\alpha_i^{\tau} = 0$, $i\neq k$, $i=1,\dots,P$. By considering $(u_1^{pk}, u_2^{pk}) = (u_1^{ek}, u_2^{ek}) + 2 \sum_{j=1}^{Q} \beta_j^{\tau}(v_1^{rj}, v_2^{rj})$, we can proceed analogously and prove that $J_1(u_1^{ek}, u_2^{ek}) = J_1(u_1^{\tau}, u_2^{\tau})$, so that in this case also a Stackelberg solution exists at an extreme point of FS. □

Remark

An immediate consequence of the preceding result is that the Stackelberg solution is an extreme point of the feasible set if it is unique.

Lemma 1.

Given any extreme point $(u_1^0,u_2^0) \in FS$, there is a neighbourhood of the point such that any elements of FS in this neighbourhood can be expressed as a convex combination of (u_1^0,u_2^0), points different from (u_1^0,u_2^0) on the extreme rays of FS originating from (u_1^0,u_2^0) and strictly convex combination of (u_1^0,u_2^0) and the extreme points of FS adjacent to (u_1^0,u_2^0).

Proof Let (v_1^0,v_2^0) be a feasible point different from (u_1^0,u_2^0). In case there is no such point, the result holds trivially. Now it is possible to find a supporting hyperplane of FS through (u_1^0,u_2^0) which has no other point in common with FS. Thus, there exist vectors d_1, d_2 and scalar f such that

$$\forall\ (u_1,u_2) \in FS \Rightarrow d_1'u_1 + d_2'u_2 \geq f$$

$$\text{and } d_1'u_1 + d_2'u_2 = f,\ (u_1,u_2) \in FS \Rightarrow (u_1,u_2) = (u_1^0,u_2^0). \tag{6.14}$$

For any $\varepsilon > 0$, define

$$FS^0(\varepsilon) = FS \cap \{(u_1,u_2)\,|\,d_1'u_1+d_2'u_2 \leq f+\varepsilon\} \tag{6.15}$$

From assumption 1, this is a nonempty, bounded convex polyhedron. Now for any ε such that $0 < \varepsilon \leq \varepsilon_1 = \|(v_1^0,v_2^0)-(u_1^0,u_2^0)\|$,

$$H^0(\varepsilon) = FS \cap \{(u_1,u_2)\,|\,d_1'u_1+d_2'u_2 \leq d_1'v_1^0+d_2'v_2^0 \text{ and}$$
$$\|(u_1,u_2)-(u_1^0,u_2^0)\| \geq \varepsilon\}$$

is a compact set containing (v_1^0,v_2^0), so that $d_1'u_1 + d_2'u_2$ attains its minimum, say $g(\varepsilon)$, on $H^0(\varepsilon)$. But then $d_1'u_1 + d_2'u_2$ has the same minimum value $g(\varepsilon)$ on the set $FS \cap \{(u_1,u_2)\,|\,\|(u_1,u_2)-(u_1^0,u_2^0)\| \geq \varepsilon\}$ and it follows from (6.14) that $g(\varepsilon) > f$. Define $N^0(\varepsilon) = \{(u_1,u_2) \mid \|(u_1,u_2)-(u_1^0,u_2^0)\| < \varepsilon\}$. Then, for any $\delta(\varepsilon)$, $0 < \delta(\varepsilon) < g(\varepsilon) - f$,

$$FS^0(\delta(\varepsilon)) \subset FS \cap N^0(\varepsilon) \tag{6.16}$$

Similarly, given δ, we have, for any $\nu(\delta)$, $0 < \nu(\delta) \leq \delta / \|(d_1,d_2)\|$,

$$FS \cap N^0(\nu(\delta)) \subset FS^0(\delta) \tag{6.17}$$

Recall the definition of index sets (definition 6.7). Then continuity implies that there always exists $\varepsilon_2 > 0$ sufficiently small such that for any ε, $0 < \varepsilon \leq \varepsilon_2$,

$$\begin{aligned}(u_1,u_2) \in N^0(\varepsilon) \Rightarrow\ & \forall\, i \notin I^0(u_1^0,u_2^0),\ A_{1i}u_1 + A_{2i}u_2 < b_i \\ & \forall\, i \notin I^1(u_1^0,u_2^0),\ u_{1i} > 0 \text{ and} \\ & \forall\, i \notin I^2\,(u_1^0,u_2^0),\ u_{2i} > 0\end{aligned} \tag{6.18}$$

Let $\varepsilon^* = \min.\,(\varepsilon_1,\varepsilon_2)$, take $\delta^* = \delta(\varepsilon^*)$ and $\nu^* = \nu(\delta^*)$. Equations (6.15)-(6.18) imply that any $(u_1,u_2) \in FS \cap N^0(\nu^*)$ can be expressed as a convex combination of the extreme points of $FS^0(\delta^*)$ and that in all points of $FS^0(\delta^*)$ exactly those constraints are inactive which are not active at (u_1^0,u_2^0). But in that case the extreme points of $FS(\delta^*)$ are precisely (u_1^0,u_2^0) and the feasible points determined by $d_1'u_1 + d_2'u_2 = f + \delta^*$ together with the linearly independent constraints which are active at (u_1^0,u_2^0). Such a point either lies on an extreme ray of FS originating from (u_1^0,u_2^0) or is a strict convex combination of (u_1^0,u_2^0) and an adjacent extreme point of FS. Thus, all $(u_1^0,u_2^0) \in FS \cap N^0(\nu^*)$ can be expressed as

$$(u_1,u_2) = \lambda_0(u_1^0,u_2^0) + \sum_{i=1}^{p} \lambda_i\{(1-\alpha_i)(u_1^0,u_2^0)+\alpha_i(u_1^{ei},u_2^{ei})\}$$

$$+ \sum_{j=1}^{q} \mu_j\{(u_1^0,u_2^0)+\beta_j(v_1^{rj},v_2^{rj})\} \tag{6.19}$$

with

$$\lambda_0 + \sum_{i=1}^{p} \lambda_i + \sum_{j=1}^{q} \mu_j = 1, \ \lambda_0 \geq 0$$

$$\lambda_i \geq 0, \quad 0 < \alpha_i < 1, \quad (i=1,\ldots,p)$$

$$\mu_j \geq 0, \ \beta_j > 0 \qquad (j=1,\ldots,q)$$

where (u_1^{ei},e_2^{ei}) $(i=1,\ldots,p)$ are the extreme points of FS adjacent to (u_1^0,u_2^0) and (v_1^{rj},v_2^{rj}) $(j=1,\ldots,q)$ the extreme rays of FS originating from (u_1^0,u_2^0).

<u>Theorem 6.3.</u>

An extreme point $(u_1^0,u_2^0) \varepsilon$ FS is a local minimum of J_1 on RC if and only if $(u_1^0,u_2^0) \varepsilon$ RC and the following conditions hold:
a) For every extreme point (u_1^{ei},u_2^{ei}) of FS adjacent to (u_1^0,u_2^0) such that $(u_1^0,u_2^0) + \alpha((u_1^{ei},u_2^{ei})-(u_1^0,u_2^0)) \varepsilon$ RC for some α, $0 < \alpha < 1$, $J_1(u_1^{ei},u_2^{ei}) \geq J_1(u_1^0,u_2^0)$ hold;
b) For every extreme ray (v_1^{rj},v_2^{rj}) of FS originating from (u_1^0,u_2^0) such that $(u_1^0,u_2^0) + \alpha(v_1^{rj},v_2^{rj}) \varepsilon$ RC for some $\alpha > 0$, $J_1(v_1^{rj},v_2^{rj}) \geq 0$ holds.
<u>Proof</u> Assume $c_{22} \neq 0$. (u_1^0,u_2^0) is a local minimum for J_1 on RC if $(u_1^0,u_2^0) \varepsilon$ RC and there exists an $\varepsilon > 0$ such that

$$(u_1,u_2) \ \varepsilon \ RC \cap N^0(\varepsilon) \Rightarrow J_1(u_1,u_2) \geq J_1(u_1^0,u_2^0) \tag{6.20}$$

Consider a strict convex combination

$$(v_1^\lambda,v_2^\lambda) = (u_1^0,u_2^0) + \lambda((u_1^{ei},u_2^{ei}) - (u_1^0,u_2^0)), \ 0 < \lambda < 1$$

of (u_1^0,u_2^0) and an extreme point (u_1^{ei},u_2^{ei}) adjacent to it. If there is one λ with $(v_1^\lambda,v_2^\lambda) \varepsilon$ RC, then $(v_1^\lambda,v_2^\lambda) \varepsilon$ RC for all λ in (0,1), using property 2 above. Choose α sufficiently small so that $(v_1^\alpha,v_2^\alpha) \ \varepsilon \ N^0(\varepsilon) \cap$ RC and eq^n.(6.20) implies

$$J_1(u_1^0,u_2^0) \leq J_1(v_1^\alpha,v_2^\alpha) = J_1(u_1^0,u_2^0) + \alpha(J_1(u_1^{ei},u_2^{ei}) - J_1(u_1^0,u_2^0)).$$

This, in turn, implies that $J_1(u_1^{ei},u_2^{ei}) \geq J_1(u_1^0,u_2^0)$ and condition (a) must hold. Condition (b) follows similarly by considering points $(w_1^\lambda,w_2^\lambda) = (u_1^0,u_2^0) + \lambda(v_1^{rj},v_2^{rj})$ with $\lambda > 0$.
If $c_{22} = 0$, the preceding conclusion follows trivially, since RC = FS in that situation.

Next suppose $(u_1^0,u_2^0) \in$ RC and conditions (a) and (b) are satisfied. From the previous lemma, there exists a $\delta > 0$ such that every $(u_1,u_2) \in FS \cap N^0(\delta)$ can be written as

$$(u_1,u_2) = \lambda_0(u_1^0,u_2^0) + \sum_{i=1}^{p} \lambda_i\{(1-\alpha_i)(u_1^0,u_2^0)+\alpha_i(u_1^{ei},u_2^{ei})\}$$

$$+ \sum_{j=1}^{q} \mu_j\{(u_1^0,u_2^0)+\beta_j(v_1^{rj},v_2^{rj})\} \tag{6.21}$$

with

$$\lambda_0 + \sum_{i=1}^{p} \lambda_i + \sum_{j=1} \mu_j = 1,\ \lambda_0 \geq 0$$

$$\lambda_i \geq 0\ ,\quad 0 < \alpha_i < 1\ (i=1,\ldots,p)$$

and $\mu_j \geq 0$, $\beta_j > 0$ $(j=1,\ldots,q)$,
where (u_1^{ei},u_2^{ei}), $i=1,\ldots,p$, are the extreme points of FS adjacent to (u_1^0,u_2^0) and (v_1^{rj},v_2^{rj}), $j=1,\ldots,q$, are the extreme rays of FS originating from (u_1^0,u_2^0).

Take any $(u_1,u_2) \in RC \cap N^0(\delta)$. Eq^n.(6.21) implies

$$J_1(u_1,u_2) = J_1(u_1^0,u_2^0) + \sum_{i=1}^{p} \lambda_i\alpha_i(J_1(u_1^{ei},u_2^{ei}) - J_1(u_1^0,u_2^0))$$

$$+ \sum_{j=1}^{q} \mu_j\beta_j J_1(v_1^{rj},v_2^{rj}) \tag{6.22}$$

Again, the case $c_{22} = 0$ is trivial. Assume that $c_{22} \neq 0$. By repeated application of property 4 to eq^n.(6.21) establishes that

$$\lambda_i > 0 \Rightarrow (1-\alpha_i)(u_1^0,u_2^0) + \alpha_i(u_1^{ei},u_2^{ei}) \in RC \text{ and}$$
$$\mu_j > 0 \Rightarrow (u_1^0,u_2^0) + \beta_j(v_1^{rj},v_2^{rj}) \in RC.$$

From conditions (a) and (b),

$$\lambda_i > 0 \Rightarrow J_1(u_1^{ei}, u_2^{ei}) \geq J_1(u_1^0, u_2^0) \text{ and}$$

$$\mu_j > 0 \Rightarrow J_1(v_1^{rj}, v_2^{rj}) \geq 0.$$

It follows immediately from eq^n.(6.22) that $J_1(u_1,u_2) \geq J_1(u_1^0,u_2^0)$ for all $(u_1,u_2) \in RC \cap N^0(\delta)$ implying that (u_1^0,u_2^0) is a local minimum of J_1 on RC. □

Corollary

An extreme point (u_1^0,u_2^0) of FS which lies in RC is a local minimum point for J_1 on RC if there are no extreme rays of FS originating from (u_1^0,u_2^0) along which J_1 decreases and all adjacent extreme points of FS yielding a lower value of J_1 are not in RC.

Proof Obvious.

4. A SIMPLEX-TYPE ALGORITHM

We follow Moraal (1981). Assume that there is always a team solution for the leader so that, by Theorem 6.1, a Stackelberg solution always exists. Theorem 6.2 immediately suggests a simplex-type algorithm to solve our present problem. We may recall the idea behind the standard simplex algorithm for LP-problem which consists of moving from extreme points of FS to adjacent extreme points while not increasing the value of the criterion and keeping all necessary informations in tableau format by means of the pivot operations. In our present problem, we may use similar steps with the only difference that, for each new point obtained in the iterative process, we have to check whether that point lies in the reaction curve or not.

One straightforward way of checking whether a feasible point (u_1^0,u_2^0) lies on the reaction curve is to substitute u_1^0 in (P1) and solve the resulting standard LP-problem. If u_2^0 is, indeed, a solution to this problem, our check is complete. But this means starting from scratch every time we complete an iteration to verify that the point obtained indeed lies on RC. The amount of computation increases tremendously if we have to continue many iteration steps.

One way out of this computational difficulty is to solve an equivalent LP-problem which is obtained by introducing artificial variables $a_1 = u_1 - u_1^0 \geq 0$ and revising the objective of the follower:

$$\min. \; J_2'(u_1,u_2,a) = c_{21}'u_1 + c_{22}'u_2 + M'a_1$$

subject to the constraints (Pl')

$$A_1u_1 + A_2u_2 \leq b, \; u_1 \geq 0, \; u_2 \geq 0$$

$$a_1 = u_1 - u_1^0, \; a_1 \geq 0$$

where $M \in \mathbb{R}^{m_1}$ and $M_i = N >> 0$, $i=1,\ldots,m_1$ (that is, the components of M are very very large). If FS is not empty, (Pl) and (Pl') are equivalent problems and if (Pl') has an optimal solution, the corresponding value of a_1 is zero; or, equivalently, the corresponding $u_1 = u_1^0$.

In checking whether a new point obtained during the iterative process indeed lies on RC, we want to use previous informations so as to reduce the computational time. We achieve this in the following manner. We work with a m x (m_1+m_2) tableau as in the simplex method and only two rows corresponding to the criteria, one for J_1 and the other for J_2. If we want to check whether a point (u_1^0,u_2^0) is on RC or not, we simply add m_1 rows to the corresponding tableau (the rows corresponding to $a_1=u_1-u_1^0$), change the row for J_2 to that for J_2' by adding $N >> 0$ times each of the new rows to it and minimize J_2' using the simplex method applied to the larger tableau.

At first sight, the above procedure seems to contradict our purpose of reducing the computational time while determining whether a point lies on RC or not. In fact, the dimension of the original problem has been further increased in the modified one. The real advantage of the above procedure becomes apparent when one tries to determine a "better" point for J_1 in RC from the optimal tableau for J_2'. The artificials may now be considered as free variables (that is, not sign restricted), since the leader determines u_1. So we change J_2' into J_2 again, delete the rows corresponding to basic artificials and consider the nonbasic artificials (if any) in the final tableau. Our assumption on the existence of a team solution for the leader allows us to pivot all these nonbasic artificials one-by-one against a basic nonartificial such that J_1 does not increase. After pivoting all nonbasic artificials to the basic and then deleting the corresponding rows, we find an extreme point of FS with a nonincreased J_1-value (and a tableau of the original size again).

It can be shown that this extreme point automatically lies on RC and it is not necessary to check this. We, therefore, can look for a better adjacent extreme point for J_1 by means of pivotsteps immediately. If such an adjacent extreme point is not in RC and the element of RC we find in this checking operation does not yield a lower value for J_1 than what we have obtained for points on RC so far, we return to the latter tableau and try another advantageous pivotstep.

This process terminates either with a tableau, corresponding to a team solution for the leader on RC (and then any Stackelberg solution is a team solution) or, with a tableau in which pivoting of a nonbasic variable with negative J_1-coefficient leads to an adjacent extreme point not in RC or to a different representation of the same (degenerate) point. Therefore, if this point, say (u_1^L, u_2^L), is nondegenerate, the conditions of theorem 6.3 are satisfied; that is, we have found a local minimum point for J_1 on RC.

To avoid further complications, assume that the feasible set is nondegenerate. To determine whether (u_1^L, u_2^L) is a Stackelberg solution we have to generate all the extreme points of $FS \cap \{(u_1,u_2) | J_2(u_1,u_2)=J_1(u_1,u_2)\}$ and check whether they lie on RC or not, and if they do, whether they are local minimum points. If we find an element of RC that is not a local minimum point, we determine one starting from that point in the manner described above. So, finally, we end up in a point of RC that is a local minimum, while all feasible points yielding the same value for J_1 are either not in RC or a local minimum points also. Using the fact that RC is connected (an assumption which probably always holds) we have obtained a (the) Stackelberg solution.

The obvious starting point of the algorithm is the determination of a team solution for the leader by the standard simplex method and then checking whether this point lies in RC in the manner described above. A flow-chart of the first part of the algorithm is given in figure 6.2 below:

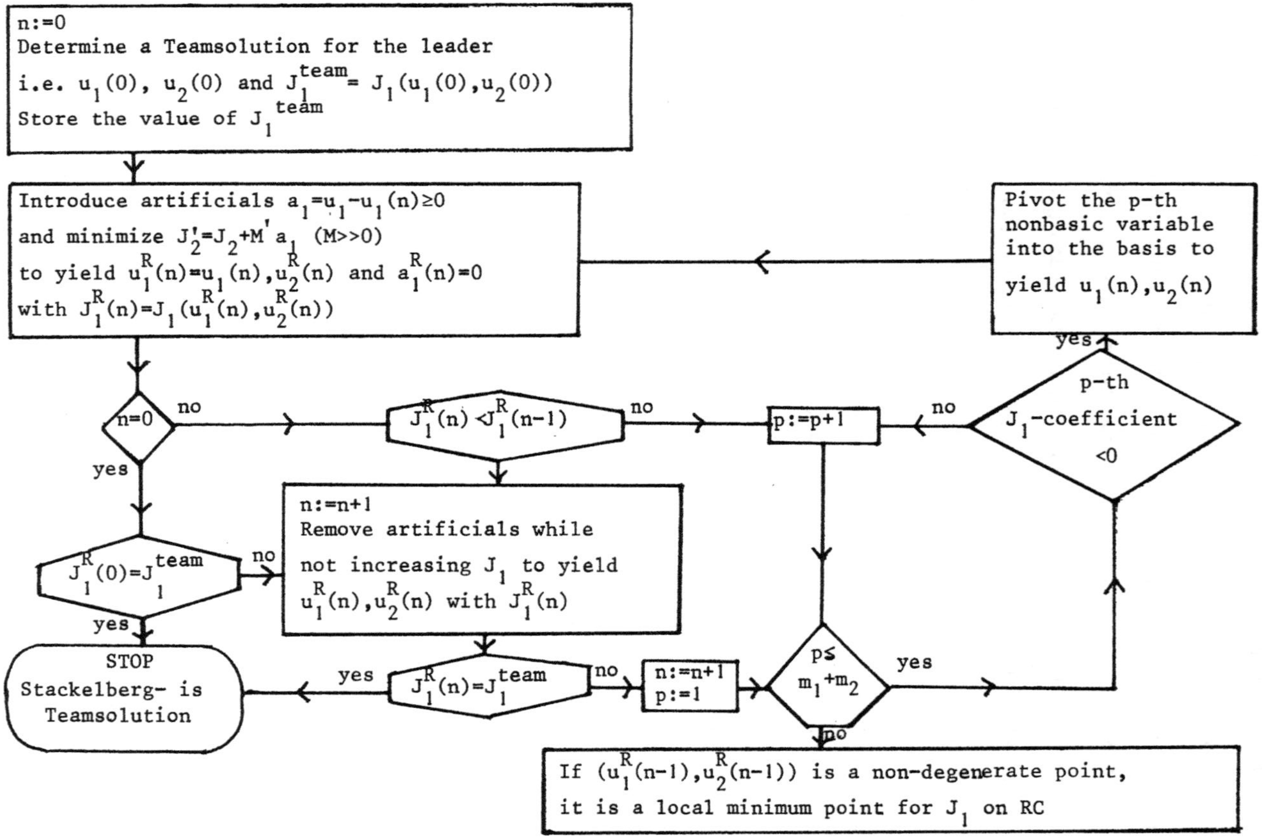

Fig. 6.2 Flow-chart to determine local minimum points for the leader on the reactioncurve.

In words, it reads as follows:

1. $n:=0$.

   Determine an optimal tableau for J_1 by means of the Simplex-method and a corresponding solution $u_1(0)$, $u_2(0)$ and memorize the optimal value of J_1, J_1^{team}. Go to step 2.
2. Add m_1 rows to the tableau from step 1 (if n=0) or step 11 (if n>0), corresponding to the artificial variables $a_1 = u_1 - u_1(n)$, add these rows N times to the J_2-row to get J_2'. Determine an optimal tableau and a corresponding solution $u_1^R(n) = u_1(n)$, $u_2^R(n)$ and $a_1^R(n) = 0$ by means of the Simplex-method. Go to step 3.
3. If n=0 then go to step 4a, else go to step 4b.

4a. If $J_1(u_1^R(0), u_2^R(0)) = J_1^{team}$ then a Stackelberg solution is a Team-solution and we stop, else go to step 5.

4b. If $J_1(u_1^R(n), u_2^R(n)) < J_1(u^R(n-1), u_2^R(n-1))$ we have found a better point on the reactioncurve: go to step 5, else we try another pivotstep from the tableau corresponding to $u_1^R(n-1), u_2^R(n-1)$ and goto step 10.

5. $n:=n+1$.

   Remove basic artificials from the tableau from step 2 and change J_2' into J_2. Pivot all nonbasic artificials into the basis against a non-artificial variable, while not increasing (preferably decreasing) the value of J_1. Remove these artificials too and we have got a tableau of the original size again, corresponding to a point $(u_1^R(n), u_2^R(n)$ on the reactioncurve. Go to step 6.
6. We have to check whether we found a Teamsolution in step 5, so if $J_1(u_1^R(n), u_2^R(n)) = J_1^{team}$ we stop and else go to step 7.
7. $n:=n+1$; $p:=1$; go to step 8.
8. If $p \le m_1 + m_2$ we have not yet checked all adjacent extreme points that can be found from the current best tableau and we go to step 9, else to step 12.
9. If the p-th coefficient in the J_1-row of the tableau corresponding

to $u_1^R(n-1), u_2^R(n-1)$ is non-negative, the corresponding adjacent extreme point is not interesting since it gives the same or a higher value to J_1 and we go to step 10, else to step 11.

10. p:=p+1. Go to step 8.
11. Copy the tableau corresponding to $u_1^R(n-1), u_2^R(n-1)$ and pivot the p-th nonbasic variable into the basis to yield a tableau corresponding to $u_1(n), u_2(n)$. Check whether this is a element of the reactioncurve by going to step 2.
12. If the point $(u_1^R(n-1), u_2^R(n-1)$ is non-degenerate, we have tried all adjacent extreme points giving a lower value to J_1, with a negative result and so this point is a local minimum point for J_1 on the reactioncurve.

The rest of the algorithm will not be given in detail, since it is a rather tedious job of bookkeeping to generate the extreme points of $FS \cap \{(u_1,u_2) | J_1(u_1^R(n-1), u_2^R(n-1))\}$, even in the simple case of non-degenaracy of FS.

If however we find another element of the reactioncurve which is not a local minimum point, we re-enter the part of the algorithm described in detail at step 6 and continue as before.

CHAPTER 7

APPLICATION TO A WORLD INDUSTRIALIZATION MODEL

1. INTRODUCTION

In the late sixties, in a U.N.I.D.O. conference held at Lima, it was agreed that by the year 2000 the industrial production of the less developed countries should be at least one-fourth the total industrial production of the world as a whole. As a follow-up to this declaration, a simple world industrialization model has been developed at Rotterdam by Opdam and Ten Kate (1978) and the feasibility of the "Lima target" has been studied within the production and trade possibilities. It was a dynamic, multisectoral, multiregional linear input-output model in which some simple linear programming exercises have been carried out. Our present study differs fundamentally from that of Opdam and Ten Kate in the role of the different regions in the optimization process. Furthermore, we assume that each region has its own optimizing criterion which may be in conflict with one another. Formulating the problem from this viewpoint leads to a multicriteria Stackelberg decision making problem in the LP-set up which has been studied in detail in the preceding chapter. For keeping the number of model constraints within reasonable bounds while retaining the basic features of the problem, we aggregate the model developed at Rotterdam and divide the world into three regions of the developed market economies, the less developed market economies and the centrally planned economies. The centrally planned economies are assumed to enter the model exogeneously. The developed and less developed economies have different decision (instrument) variables and different optimizing criteria and the developed economies have the role of a leader in the optimization process. This leads to a Stackelberg game problem mentioned above. This chapter is an expanded version of Bagchi et al (1982) and makes use of the modified model developed by Moraal (1980).

2. A WORLD INDUSTRIALIZATION MODEL

We start with a very general model description where there are arbitrary regions and time periods. Region and time will be denoted by lower indices r and t, respectively. We start with time index zero. Variable vectors are denoted by small Roman letters, scalar variables by capital Roman letters, coefficients and vectors of coefficients are denoted by small Greek letters, matrices of coefficients by capital Roman letters with an upper bar and diagonal matrices of coefficients by small Greek letters with a circumflex.

We want to develop an interregional dynamic input-output model.

Balance Equations

Each region consists of several sectors indicating different economic activities. In any region at any time, sectoral gross output at least equals sectoral demand and net foreign sectoral demand, measured by some common standard (in application, we value all flows at 1970 market prices f.o.b.*) :

$$x_{rt} \geq \bar{A}_{rt}\, x_{rt} + c_{rt} + i_{rt} + e_{rt} - m_{rt} \qquad (7.1)$$

where x = vector of gross output by sector;

$\bar{A}$ = technology matrix of input-output coefficients;

c = vector of final consumption by sector;

i = vector of total investments by sector of origin;

e = vector of exports by sector;

m = vector of imports by sector.

Consumption, savings and income

Gross domestic product by region is defined as the sum over sectors of gross value added by sector:

$$Y_{rt} = \alpha_{rt}'\, x_{rt} \qquad (7.2)$$

---

*f.o.b. is the abbreviation for "free on board"

where

Y = gross domestic product by region

α = vector of gross value added coefficients

and the prime denotes transpose.

For the consumer's demand, a simple Engel curve is used:

$$c_{rt} = \gamma_{0rt} P_{rt} + \gamma_{1rt} C_{rt} \qquad (7.3)$$

where P = total population by region;

C = total consumption expenditure by region;

γ_0 = vector of coefficients, satisfying the condition that the sum of its elements is equal to zero;

γ_1 = vector of coefficients, satisfying the condition that the sum of its elements equals one.

Savings consists of two parts. The endogeneous part of total savings may be influenced by means of Government action. We denote this part of savings by S_{rt} and consider it as instrumental or control variable. The autonomous part of savings, $\sigma_{0rt} P_{rt}$, is predetermined. Then the total consumption by region is given by

$$C_{rt} = Y_{rt} - S_{rt} - \sigma_{0rt} P_{rt} \qquad (7.4)$$

where

S = controllable part of savings (in the original model $S_{rt} = \sigma_{1rt} Y_{rt}$)

σ_0 = a scalar coefficient, expressing the autonomous part of per capita savings.

Investments and Capital Formation

Investments are used for replacing worn out capital stocks and also for new investments for capital formation. We, therefore, write

$$i_{rt} = i_{rt}^{repl.} + i_{rt}^{new} \qquad (7.5)$$

where

$i^{repl.}$ = vector of replacement investments by sector of origin;

i^{new} = vector of new investment by sector of origin.

We first try to determine new investments by sector of destination. We assume that new investments by sector of destination over a period of θ years are proportional to the increase of that sector's gross output over this period. One time step in our model will mean a period of θ years. Then

$$h_{rt} = \hat{\kappa}_{rt}(x_{rt}-x_{r\overline{t-1}}) \tag{7.6}$$

where

h = vector of accumulated new investments over a period of θ years by sector of destination;

$\hat{\kappa}$ = diagonal matrix of marginal sectoral capital-output ratios.

To arrive at new investments by sector of origin, the distribution of new investment deliveries over new investment requirements has to be specified. This is given by means of a distribution matrix $\bar{D}$. Thus, defining $\bar{K} = \bar{D}\hat{\kappa}$, we see that $\bar{K}$ is the matrix of partial marginal sectoral capital-output ratios, and its (i,j)-th element $\bar{K}(i,j)$ denotes the quantity of new investment goods delivered by sector i to sector j that is necessary per unit increase of production in sector j. This helps us to translate new investment requirements by sector of destination into new investment deliveries by sector of origin over a period of θ years. We have to obtain an expression for new investment in the terminal year which will, of course, only be a fraction of new investments in producing sectors over a period of θ years. This fraction, usually referred to as a "stock-flow conversion factor", will be defined implicitly. The average yearly new investments by the delivering sectors over a period of θ years; namely, $\frac{1}{\theta}\bar{D}k$, is taken to be a weighted sum of new investments by delivering sectors in the initial and terminal year of the period under consideration which can be mathematically expressed as

$$\hat{\omega}_{0rt}\, i^{new}_{r\overline{t-1}} + \hat{\omega}_{1rt}\, i^{new}_{rt} = \frac{1}{\theta}\bar{D}_{rt}\, h_{rt} \tag{7.7}$$

where $\hat{\omega}_0$, $\hat{\omega}_1$ = diagonal matrices of weights, satisfying $\hat{\omega}_0 + \hat{\omega}_1 = I$, the identity matrix.

For further details on stock-flow conversion factor, we refer the reader to Appedix II of the report of Opdam and Ten Kate (1978). Equation (7.7) may be rewritten as

$$i_{rt}^{new} = \hat{\omega}_{1rt}^{-1}(\tfrac{1}{\theta}\bar{D}_{rt}h_{rt} - \hat{\omega}_{0rt}i_{r\overline{t-1}}^{new}) \qquad (7.7')$$

We have yet to quantify the replacement investments. We assume that they are proportional to the capital stocks existing θ years before, so that

$$i_{rt}^{repl.} = \hat{\delta}_{0rt}\, k_{r\overline{t-1}} \qquad (7.8)$$

where

k = vector of capital stocks of homogeneous capital goods;

$\hat{\delta}_0$ = diagonal matrix of replacement ratios.

In the starting year of our model, we take

$$i_{r0}^{repl.} = \hat{\delta}_{1r}k_{r0} \qquad (7.8a)$$

where $\hat{\delta}_1$ = diagonal matrix of replacement ratios (the elements of $\hat{\delta}_1$ will be smaller than the corresponding elements of $\hat{\delta}_0$).

Finally, to specify capital stocks of homogeneous capital goods, we assume that marginal capital-output ratios equal average capital-output ratios. Thus, we have

$$k_{rt} = \bar{K}_{rt}\, x_{rt} \qquad (7.9)$$

Trade

For bringing trade into the model, we make a distinction between sectors which produce tradeable (or world) goods and sectors which produce non-tradeable (or regional) goods. We can always order the sectors in such a way that all sectors which produce world goods correspond to the first part of the components of all the vector valued variables mentioned so far. The remaining components then correspond to sectors which produce only regional goods. Trade variables for regional goods producing sectors are clearly zero, except for the SERVICES sector as is explained below.

At the world level, total imports by sector equal total exports by sector if flows are measured by the same units. As mentioned already, all flows in the model are valued at 1970 world market prices f.o.b. However, c.i.f.* margins have to be paid by the importing regions and are received by the regions providing the insurance-freight services. Therefore, these margins appear as imports and/or exports of the regional goods sector which in the model we refer to as the SERVICES sector.

We assume that the margins on imports of world goods of certain sectors are exported by the region that imports these world goods and that the margins on imports of the other world goods producing sectors are exported by the region that exports goods of these sectors.

The above description leads to the following equations:

$$m^s_{rt} = \phi'_t \, m^w_{rt}$$

$$e^s_{rt} = \phi'_t \hat{\xi} \, m^w_{rt} + \phi'_t (1-\hat{\xi}) e^w_{rt}$$

$$m^n_{rt} = 0$$

$$e^n_{rt} = 0$$

where

m^s, e^s = scalars denoting imports and exports of the SERVICES sector, respectively;

m^n, e^n = vectors of imports and exports of sectors producing regional goods, except the SERVICES sector, respectively;

m^w, e^w = vectors of imports and exports of sectors producing world goods, respectively;

ϕ = vectors of c.i.f. margins on imports by sector;

$\hat{\xi}$ = diagonal matrix with unit or zero entries on the main diago-

---

*c.i.f. is the abbreviation for "cost, insurance and freight".

nal (unit and zero entries corresponding to sectors for which c.i.f. margins are exported by importing regions and exporting regions; respectively).

As a result of the f.o.b. valuation of the model's variables, imports by the SERVICES sector vanish from the regional balance equations. They, however, play a role in the balance of trade definition. Summarizing, we obtain the following equations for exports and imports in the regional balance equations:

$$m_{rt} = \bar{B}_1 m^w_{rt} \qquad (7.10)$$

$$e_{rt} = \bar{B}_{2t} e^w_{rt} + \bar{B}_{3t} m^w_{rt} \qquad (7.11)$$

where

$$\bar{B}_1 = \begin{bmatrix} I \\ 0_n \\ 0 \end{bmatrix}, \quad \bar{B}_2 = \begin{bmatrix} I \\ 0_n \\ \phi'(I-\hat{\xi}) \end{bmatrix} \text{ and } \bar{B}_3 = \begin{bmatrix} I \\ 0_n \\ \phi'\hat{\xi} \end{bmatrix},$$

I and 0 being the identity and zero matrices, respectively, corresponding to the number of sectors producing world goods and 0_n is a zero matrix, the size of which corresponds to the number of regional goods minus one (the one corresponding to the SERVICES sector).
It is rather easy to influence imports and exports by Governmental action (e.g., by means of import duties and/or import restrictions) and therefore, some trade variables may be chosen as control or instrumental variables.
It is easier to control one's own imports unilaterally than exports, because exports are also influenced by the other regions. Therefore, exports of world goods are determined endogeneously. This is done by distinguishing imports of world good by region of origin:

$$m^w_{rt} = \sum_{i=1}^{R} m^{wi}_{rt} \qquad (7.12)$$

where m^{wi} = vector of imports of world goods imported from region i;
R = number of regions in the model.

Since imports of region r from region i equal exports from region i to region r, we obtain the following equation for exports of world goods:

$$e^{w}_{rt} = \sum_{i=1}^{R} m^{wr}_{it} \tag{7.13}$$

In equations (7.12) and (7.13), the trade variables explicitly stand for total trade; that is, interregional and also intraregional trade. Furthermore, these equations imply that at the world level, exports of world goods by sector equal imports of world goods by sector, which can easily be verified.

For each region, the imports of world goods imported from the other regions are considered to be controllable; that is, for region r the vectors m^{wi}_{rt} for $i \neq r$ are control or instrumental vectors. The only influence of trade on the model's outcome is via the c.i.f. margins, because imports of c.i.f. margins disappear from the balance equations. If we, therefore, took intraregional trade as a control or instrumental variable, a region might artificially change its gross output, for example, by unnecessarily transporting goods to and fro inside the region. We consider intraregional trade as exogeneously determined and do not take them as control or instrumental variables.

A number of additional constraints have to be imposed to make the problem realistic. We must have balance of trade restrictions, import substitution limits, export growth limitations, savings constraints and finally, of course, the constraint arising out of the Lima target. We consider them in the next section in connection with our specific aggregated model. The optimizing criteria will also be introduced then.

Equations (7.1)-(7.13) describe a general input-output model of a world industrialization model consisting of several regions with several sectors of economic activities in each region.

3. A SIMPLE AGGREGATED MODEL

In the model developed by Opdam and Ten Kate (1978), the world was divided into three regions; namely, the developed market economies

(DMEs, denoted by region index 1), the less developed market economies (LMEs, denoted by region index 2) and the centrally planned economies (CPEs, denoted by region index 3).

Furthermore, production activity in each region was divided into nine sectors: Agriculture, mining, food, chemicals, metals, equipment, transport and communication, construction, services. The first six of these sectors were considered to produce world goods; that is, goods tradeable on a world-wide scale. The remaining three sectors were considered to produce only regional goods, with the exception that c.i.f. margins on imports were received by the SERVICES sector. It was assumed that c.i.f. margins on imports of sectors producing agricultural, mining and food products were exported by the importing region, whereas for sectors producing chemicals, metals and equipment, c.i.f. margins were exported by the exporting region.

For the SLP-problem of our interest, even this aggregation leads to an unwieldy optimization problem with too large a number of state and decision variables. In our present study, we consider a still simpler aggregated model where we consider only three sectors:

1. The "TRADITIONAL" sector, composed of sectors producing agricultural, mining and food products (denoted by index 1).
2. The "(MODERN) INDUSTRY" sector, composed of sectors producing chemicals, metals and equipment (denoted by index 2).
3. The "SERVICES" sector, consisting of transport, communication, construction and services (denoted by index 3).

According to the original classification, the traditional and industrial sectors produce tradeable world goods, whereas the services sector produces regional goods. However, c.i.f. margins will be accounted for by means of exports and imports of the SERVICES sector. C.i.f. margins on imports are exported by importing regions for goods from the traditional sector and by exporting regions for goods from the modern industrial sector. As mentioned already, all flows are valued at 1970 market prices f.o.b. and therefore, imports of c.i.f. margins by the SERVICES

sector disappear from the regional balance equations.
For easy reference we rewrite equation (7.2)-(7.13) once again. The balance equation is given by

$$x_{rt} \geq \bar{A}_r x_{rt} + c_{rt} + i_{rt} + e_{rt} - m_{rt}$$

where x stands for gross output, c is final consumption, i is total investment by sector of origin, e stands for exports and m for imports.
Per capita consumption is specified by a simple Engel curve

$$c_{rt} = \gamma_{0r} P_{rt} + \gamma_{1r} C_{rt}$$

where P stands for total population and C, the total consumption expenditure, is given by

$$C_{rt} = Y_{rt} - S_{rt} - \sigma_{0r} P_{rt}.$$

Here Y is the gross domestic product satisfying

$$Y_{rt} = \alpha'_{rt} x_{rt}$$

and S denotes the controllable part of the savings.
Turnings to investments and capital formations, we have

$$i_{rt} = i_{rt}^{repl.} + i_{rt}^{new}$$

where $i^{repl.}$ stands for replacement investments and i^{new} for new investments.

$$i_{rt}^{repl.} = \hat{\delta}_{0r} k_{r\,\overline{t-1}} \quad \text{(k:capital stocks)}$$

$$k_{rt} = \bar{K}_r x_{rt}$$

$$i_{rt}^{new} = \hat{\delta}_1^{-1}(\tfrac{1}{\theta}\bar{D}_r k_{rt} - \hat{\omega}_0\, i_{r\,\overline{t-1}}^{new})$$

$$k_{rt} = \hat{\kappa}_r (x_{rt} - x_{r\overline{t-1}})$$

k is the accumulated new investments over a period of θ years by sector of destination. θ is 10 years in the model. In our present analysis, we assume that the centrally planned economies enter the model only exogeneously. The equations above, therefore, are of interest to us only for r = 1,2. The region r = 3 enter the model only through the trade equations. m^w and e^w denote imports and exports of sectors producing world goods, while m^{wi} denotes imports of world goods imported from region i. With these conventions, we have

$$e_{rt} = B_2\, e^w_{rt} + B_3\, m^w_{rt}$$

$$m_{rt} = B_1\, m^w_{rt}$$

$$e^w_{rt} = \sum_{i=1}^{3} m^{wr}_{it}, \quad m^w_{rt} = \sum_{i=1}^{3} m^{wi}_{rt}.$$

This completes our description of the input-output model. Let us define

$$R_r = I - \bar{A}_r - \gamma_{1r}\alpha'_r - \hat{\omega}_1^{-1}\,\frac{1}{\theta}\bar{K}_r.$$

Then the above set of equations give two basic constraints for the model:

$$R_r x_{rt} \geq (\hat{\delta}_{0r} - \hat{\omega}_1^{-1}\,\frac{1}{\theta})\bar{K}_r x_{r\overline{t-1}} - \hat{\omega}_1^{-1}\hat{\omega}_0\, i^{new}_{r\,\overline{t-1}}$$
$$+ (B_3 - B_1)m^{wi}_{rt} - \gamma_{1r}S_{rt} + B_2 m^{wr}_{it} + (\gamma_{0r} - \gamma_{1r}\sigma_{0r})P_{rt} \qquad (7.14)$$

and

$$i^{new}_{rt} = \hat{\omega}_1^{-1}\,\frac{1}{\theta}\,\bar{K}_r x_{rt} - \hat{\omega}_1^{-1}\,\frac{1}{\theta}\,\bar{K}_r x_{r\overline{t-1}} - \hat{\omega}_1^{-1}\hat{\omega}_0\, i^{new}_{r\overline{t-1}} \qquad (7.15)$$

$$r = 1,2;\; i = 1,2;\; i \neq r,\; t = 1,2,3.$$

We take time steps of 10 years with 1970 as the base. Thus, the terminal year for our model is the year 2000.

We now specify the additional constraints mentioned towards the end of the preceding section. We denote by x_{rtj} the j-th component of the vector x_{rt} and use similar notations for other vectors appearing in the model. There must be some allowable limits to balance of trade deficits and surpluses. We assume that the CPEs have perfect trade balance with other regions. Then we can express balance of trade restrictions as

$$B^{\ell}_{rt} \leq [1 \quad 1](m^{wr}_{it} - m^{wi}_{rt}) \leq B^u_{rt} \qquad (7.16)$$

There is also a limit to import substitution. We represent this by

$$m^{wi}_{rtj} \geq \{(1-\varepsilon^i_{rj})^t\, m^{wi}_{r0j}/x_{r0j}\}x_{rtj}$$

with ε^i_{rj}, $i \neq r$, $i,j,r = 1,2$, constants. We take, as in the original model, $\varepsilon^2_{11} = \varepsilon^2_{12} = \varepsilon^1_{21} = \varepsilon^1_{22} = 0.2$. Given the initial export levels, there is cearly <u>limit to export growth</u> for world goods by region of destination which we take to be of the geometric type:

$$m^{wr}_{it} \leq (1+\hat{\pi}^i_r)^{t.\theta} \; m^{wr}_{i0} \tag{7.17}$$

There are also <u>savings constraints</u>. We assume that savings cannot exceed one-fourth the national output, which is a quite plausible assumption. This gives us

$$S_{rt} \leq 0.25 \; Y_{rt} = 0.25 \; \alpha'_r \; x_{rt} \tag{7.18}$$

Last, but not the least, is the constraint imposed on the model by the "Lima Target". This means that, by the year 2000, we would like the industrial production of the less developed countries to be at least one-fourth the total industrial production of the world as a whole. In our notation, we can express this target mathematically as

$$x_{232} \geq \alpha(x_{132}+x_{232})$$
$$\alpha = 0.25(x_{102}+x_{202}+x_{302})/(x_{102}+x_{202}) \tag{7.19}$$

Equations (7.14)-(7.19) constitute the model constraints. There are altogether 51 inequalities in this model. Control variables for regions r, r = 1,2 are x_{rt}, i^{new}_{rt}, m^{wi}_{rt}, S_{rt}, $i = 1,2$, $i \neq r$, $t = 1,2,3$. Both DMEs and LMEs have altogether 27 control or instrumental variables. Our next task is to specify the optimizing criteria of the DMEs and LMEs. This is a complicated question. We consider here a linear criterion for both the decision makers. The simple criterion of discounted total consumption will be used. Thus, for r = 1,2, region r wants to <u>minimize</u>.

$$J_r = - \sum_{t=1}^{3} C_{rt}/(1+\eta_r)^{\theta(t-1)} \; , \; \eta_r \text{ some specified constant.}$$

Using the consumption equation (7.4) and choosing η_r, r = 1,2, such that $(1+\eta_r)^\theta = 2$, we can write

$$J_r = \{\sigma_{0r}P_{r1} + \tfrac{1}{2}\sigma_{0r}P_{r2} + \tfrac{1}{4}\sigma_{0r} \; P_{r3}\} + \tag{7.20}$$

$$+ \{-\alpha_r' x_{r1} + S_{r1} - \tfrac{1}{2}\alpha_r' x_{r2} + \tfrac{1}{2}S_{r2} - \tfrac{1}{4}\alpha_r' x_{r3} + \tfrac{1}{4}S_{r3}\}$$

We assume that the DMEs have the role of a leader in our game-theoretic model. Given the control variables of the DMEs (r=1), the LMEs (r=2) determine their controls so that J_2 is minimized. Given those optimum control variables for LMEs and substituted in the criterion J_2, DMEs determine their control variables so that J_1 is a minimum.
We are, therefore, faced with a Stackelberg Linear Programming problem studied in the preceding chapter.

4. NUMERICAL STUDIES

For easy reference, we give a compact overview of all the notations that have been used in building the model. Thus,

- x: vector of gross output by sector
- $\bar{A}$: technology matrix of input-output coefficients
- c: vector of final consumption by sector
- i: vector of total investments by sector of origin
- e: vector of exports by sector
- m: vector of imports by sector
- Y: gross domestic product by region
- α: vector of gross value added coefficients
- P: total population by region
- C: total consumption expenditures by region
- γ_0: vector of coefficients with the sum of its elements equal to zero
- S: controllable part of savings
- σ_0: scalar coefficient giving autonomous part of savings per capita
- $i^{repl.}$: vector of replacement investments by sector of origin
- i^{new} : vector of new investments by sector of origin
- h: vector of accumulated new investments over a period of θ years by sector of destination
- $\hat{\kappa}$: diagonal matrix of marginal sectoral capital-output ratios

$\hat{\omega}_0, \hat{\omega}_1$: diagonal matrix of weights; $\hat{\omega}_0 + \hat{\omega}_1 = I$

k: vector of capital stocks

$\hat{\delta}_0$: diagonal matrix of replacement ratios

γ_1: vector of coefficients with the sum of its elements equal to one

All the coefficients appearing in equations (7.1)-(7.13) were estimated in the original model of Opdam and Ten Kate (1978). In the present model the sectors are further aggregated and this necessitates recalculations of the coefficients. This has been carried out in detail in Moraal (1980 In the original model, for the input-output table of the DMEs, a 1965 input-output table for the EEC was used.For the LMEs, two tables were estimated, one based on 1973 input-output table for India and the other based on 1960 table for Mexico. We use the Indian data. Finally, the table for CPEs was based on 1966 input-output table for the USSR.
To aggregate the tables used in the original model to conform with our present sectoral classification, we simply add the relevant figures of tables in the original model over sectors constituting a new sector. We also omit the input-output table for the CPEs as they are not directly relevant to our present study. Thus, we have the following input-output tables of interest:

	Intermediate deliveries			Final demand			
Sector	1	2	3	c	i	e	-m
1	224378	49844	74061	320100	11649	62849	- 88569
2	54020	333984	230945	137039	216245	161809	-137393
3	79028	143121	427522	1167497	250994	12166	0
Values added	294948	469688	1336173				
Groos output	654371	996664	2068711				

Table 7.1, Input-output table for the DMEs 1970 (in millions of U.S. dollars)

Source: Table 2.13 in Opdam en Ten Kate (1978).

	Intermediate deliveries			Final demand			
Sector	1	2	3	c	i	e	-m
1	44666	10126	4863	162755	5632	43852	-18091
2	18021	29205	18844	16967	38328	11448	-35810
3	37889	18107	35376	125448	33225	1880	0
Value added	153228	39566	190450				
Gross output	253819	97018	249547				

Table 7.2, Input-output table for the LMEs 1970 (in millions of U.S. dollars) based upon Indian table.

Source: Table 2.14 in Opdam and Ten Kate (1978).

From tables 7.1 and 7.2, base year values of technical and value added coefficients can be simply computed by dividing the entries of the left hand side of these tables by sectoral gross outputs. This leads to the following matrices of technical coefficients and vectors of value added coefficients:

$$\bar{A}_1 = \begin{bmatrix} 0.3429 & 0.0500 & 0.0358 \\ 0.0826 & 0.3351 & 0.1116 \\ 0.1208 & 0.1436 & 0.2067 \end{bmatrix}$$

$$\bar{A}_2 = \begin{bmatrix} 0.1760 & 0.1044 & 0.0195 \\ 0.0710 & 0.3010 & 0.0755 \\ 0.1493 & 0.1866 & 0.1418 \end{bmatrix} \qquad (7.21)$$

$$\alpha_1' = [0.4538 \quad 0.4713 \quad 0.6459]$$
$$\alpha_2' = [0.6037 \quad 0.4078 \quad 0.7632] \qquad (7.22)$$

We next determine the consumption coefficients appearing in eq^n (7.3). For the DMEs and LMEs, we assume that the same consumption relation holds; that is, the vectors of coefficients γ_0 and γ_1 are the same for the DMEs and the LMEs. The 1970 populations of the DMEs and LMEs were approximately 710 and 1725 million people, respectively. Now we determine the intercepts and slopes for the DMEs and LMEs in the base year from the consumption figures in tables 7.1 and 7.2 and the population figures mentioned above, assuming that $\gamma_{01} = \gamma_{02}$ and $\gamma_{11} = \gamma_{12}$. This gives us

$$\begin{aligned} \gamma_{11}' &= \gamma_{12}' = [0.1688 \quad 0.0868 \quad 0.7444] \\ \gamma_{01}' &= \gamma_{02}' = [64.48 \quad -5.51 \quad -58.97] \end{aligned} \qquad (7.23)$$

Autonomous savings per capita is estimated in the same way as consumption coefficients have been estimated. It is assumed that the same savings function hold for the DMEs and LMEs and that autonomous savings per capita is the same in both the regions. This gives us

$$\sigma_{01} = \sigma_{02} = 5.68 \qquad (7.24)$$

Our next task is to determine coefficients arising in models for investment and capital formation. For this, we have to aggregate capital-output tables given in the original model, where dynamically consistent capital output matrices were given for the base year (tables 2.17, 2.18, 2.19 and 2.21 in Opdam and Ten Kate (1978)). Dynamic consistency means that regional investment levels by sector of origin in the base year are compatible with both the required replacement of existing capital stocks and the achieved sectoral growth rates over the θ year period inmediately preceding the base year (θ=10 years both in the original and the present model). Tables of capital stocks by sector of origin and sector of destination in the base year are computed from the figures given in tables 2.17, 2.18 and 2.21 of the original model of Opdam and Ten Kate (1978) and the gross outputs of sectors in 1970 of the original model. These tables are then aggregated by adding over sectors constituting a sector in the present model. Furthermore, replacement investment by new sectors are determined by adding replacement investment figures for the original model corresponding to appropriate sectors. Using the already determined aggregated total investments, new investments by sector of origin in our present classification are determined by the difference between total investments by sector of origin and replacement investments by sector of origin. The results, along with the aggregated gross output figures are given below:

Origin \ Destination	1	2	3	k	i^{repl}	i^{new}	i
1	136568	0	0	136568	6146	5503	11649
2	320645	763439	1116260	2200344	99015	117230	216245
3	304519	351141	2393763	3049423	96050	154944	250994
Gross output	654371	996664	2068711				

Table 7.3, Capital and investment figures for the DMEs 1970 (in millions of U.S. dollars).

Source: Table 2.17 and Table 2.21 in Opdam and Ten Kate (1978)

Origin \ Destination	1	2	3	k	i^{repl}	i^{new}	i
1	66240	0	0	66240	2981	2651	5632
2	98149	149592	109785	357526	16089	22239	38328
3	179588	55661	159151	394400	13062	20163	33225
Gross output	253819	97018	249547				

Table 7.4, Capital and investment figures for the LMEs 1970 (in millions of U.S. dollars), based upon Indian table.

Source:Table 2.18 and Table 2.21 in Opdam and Ten Kate (1978).

Since capital-ouput ratio in the base year for sectors i and j equals the base year capital stock needed for production in sector j and originating from sector i divided by the base year gross output in sector j, capital-output ratios for our present sectoral classification may be easily computed from the Tables 7.3 and 7.4. This enables us to determine the matrices $\bar{K}_1$ and $\bar{K}_2$. Adding the elements of the matrices $\bar{K}_1$ and $\bar{K}_2$ by columns give us the diagonal elements of the diagonal matrices $\hat{\kappa}_1$ and $\hat{\kappa}_2$. Finally, the distribution matrices $\bar{D}_1$ and $\bar{D}_2$ are determined by the relation $\bar{D}_r\hat{\kappa}_r = \bar{K}_r$, r = 1,2. This gives us

$$\bar{K}_1 = \begin{bmatrix} 0.2087 & 0 & 0 \\ 0.4900 & 0.7660 & 0.5396 \\ 0.4654 & 0.3523 & 1.1571 \end{bmatrix} \tag{7.25a}$$

$$\hat{\kappa}_1 = \text{diag. } [1.1641 \quad 1.1183 \quad 1.6967] \tag{7.26a}$$

$$\bar{D}_1 = \begin{bmatrix} 0.1793 & 0 & 0 \\ 0.4209 & 0.6850 & 0.3180 \\ 0.3998 & 0.3150 & 0.6820 \end{bmatrix} \tag{7.27a}$$

$$\bar{K}_2 = \begin{bmatrix} 0.2610 & 0 & 0 \\ 0.3867 & 1.5419 & 0.4399 \\ 0.7075 & 0.5737 & 0.6378 \end{bmatrix} \tag{7.25b}$$

$$\hat{\kappa}_2 = \text{diag. } [1.3552 \quad 2.1156 \quad 1.0777] \tag{7.26b}$$

$$\bar{D}_2 = \begin{bmatrix} 0.1926 & 0 & 0 \\ 0.2853 & 0.7288 & 0.4082 \\ 0.5221 & 0.2712 & 0.5918 \end{bmatrix} \tag{7.27b}$$

In tables 7.3 and 7.4, replacement investment by sector of origin and existing capital stocks by sector of origin in the base year are given. Now, replacement ratios for the base year related to the current capital stocks (the elements of the diagonal matrices $\hat{\delta}_{1r}$) are determined by means of the relation $i_{r0}^{repl.} = \hat{\delta}_{1r0}k_{r0}$ and the figures given in tables 7.3 and 7.4. This gives us

$$\hat{\delta}_{11} = \text{diag. } [0.0450 \quad 0.4550 \quad 0.0315]$$
$$\hat{\delta}_{12} = \text{diag. } [0.0450 \quad 0.4550 \quad 0.0331]$$

where we have suppressed the last suffix 0 indicating that all figures relate to the base year. We give replacement ratios for the base year related to capital stocks one decade before (the elements of the diagonal matrix $\hat{\delta}_{0r0}$) twice the value of the replacement ratios for the base year related to current capital stocks. This comes down to a postulated annual growth of capital stocks of approximately 7 %. Then

$$\hat{\delta}_{01} = \text{diag. } [0.0900 \quad 0.0900 \quad 0.0630]$$
$$\hat{\delta}_{02} = \text{diag. } [0.0900 \quad 0.0900 \quad 0.0662] \qquad (7.28)$$

The weighting matrices $\hat{\omega}_0$ and $\hat{\omega}_1$ are taken to be the same for all regions and are determined by

$$\hat{\omega}_0 = \omega_0 I \text{ and } \hat{\omega}_1 = \omega_1 I$$

where I is the identity matrix and

$$\omega_0 = 0.55 \text{ and } \omega_1 = 0.45 \qquad (7.29)$$

In fact, this implies a postulated annual growth of investment by 6 %. The only coefficients not yet specified are the matrices $\bar{B}_1$, $\bar{B}_2$ and $\bar{B}_3$. This needs knowing the vector ϕ of c.i.f. margin on imports by sector. Using the trade figures given in the original model (tables 2.7 and 2.8 in Opdam and Ten Kate (1978)), ϕ has been calculated for the present model in Moraal (1980) and is given by

$$\phi' = [0.09964 \quad 0.01958].$$

Equations (7.10) and (7.11) give us

$$B_1 = \begin{bmatrix} 1 & 0 \\ 0 & 1 \\ 0 & 0 \end{bmatrix}, \quad B_2 = \begin{bmatrix} 1 & 0 \\ 0 & 1 \\ 0 & 0.01958 \end{bmatrix}, \quad B_3 = \begin{bmatrix} 0 & 0 \\ 0 & 0 \\ 0.09964 & 0 \end{bmatrix} \qquad (7.30)$$

The population in millons for the regions in 1980 and 1990 are estimated to be

$$P_{11} = 784 \qquad P_{12} = 857 \text{ and } P_{13} = 928$$
$$P_{21} = 2187 \qquad P_{22} = 2745 \text{ and } P_{23} = 3412. \qquad (7.31)$$

The initial conditions, in millions of U.S. dollars, are

$$x_{10} = \begin{bmatrix} 654371 \\ 996664 \\ 2068711 \end{bmatrix} \qquad x_{20} = \begin{bmatrix} 253819 \\ 97018 \\ 249547 \end{bmatrix} \qquad x_{30} = \begin{bmatrix} 441015 \\ 453719 \\ 566269 \end{bmatrix}$$

$$B^u_{11} = - B^\ell_{11} = B^u_{21} = - B^\ell_{21} = 15000$$
$$B^u_{12} = - B^\ell_{12} = B^u_{22} = - B^\ell_{22} = 30000$$
$$B^u_{13} = - B^\ell_{13} = B^u_{23} = - B^\ell_{23} = 45000.$$

$(m_{10}^{w2})' = [43852 \quad 11448]$ and $(m_{20}^{w1})' = [18093 \quad 35807]$

The choice of α is, of course, subjective. With the initial values, "Lima" goal could make α approximately equal to 0.25 (996664+97018+453719)/(996664+97018) ≈ 0.35. But with the expected gradual increase in the share of industrial production of the L.M.E.s a more reasonable choice of α is 0.3. Simulations were performed with several choices for α. Here we shall give results only for α = 0.3 and α = 0.2 and based on these figures, we draw some general conclusions.

Team solution and the same Stackelberg solution for α = 0.2

In the following tables, the optimum values of the different variables are given (we take their 1970 values to be 1.000) and next to those quantities we give in parenthesis the corresponding yearly growth percentages (averaged over a period of 10 years).

t	x_{1t1}	x_{1t2}	x_{1t3}
1	1.504 (4.16)	1.764 (5.84)	1.701 (5.46)
2	2.655 (5.85)	2.923 (5.18)	2.890 (5.44)
3	5.062 (6.66)	4.470 (4.34)	4.818 (5.24)

1970 values: 654371 996664 2068711

t	x_{2t1}	x_{2t2}	x_{2t3}
1	1.730 (5.64)	1.417 (3.55)	2.126 (7.83)
2	2.354 (3.13)	4.199 (11.48)	3.922 (6.31)
3	2.354 (0.00)	11.481 (10.58)	7.761 (7.06)

253819 97018 249547

t	i_{1t1}^{new}	i_{1t2}^{new}	i_{1t3}^{new}
1	1.555 (4.51)	1.673 (5.28)	1.789 (5.99)
2	4.450 (11.09)	2.847 (5.46)	2.980 (5.23)
3	7.834 (5.82)	4.302 (4.21)	4.807 (4.90)

1970 values: 5503 117230 154944

t	i^{new}_{2t1}	i^{new}_{2t2}	i^{new}_{2t3}
1	2.834 (10.98)	1.352 (3.06)	2.454 (9.39)
2	0.000 (——)	5.088 (14.17)	3.092 (2.34)
3	0.000 (——)	8.877 (5.72)	7.422 (9.15)
1970 values:	2651	22239	20163

t	m^{w2}_{1t1}	m^{w2}_{1t2}	m^{w1}_{2t1}	m^{w1}_{2t2}
1	2.594 (10.00)	1.411 (3.50)	1.384 (3.31)	2.509 (9.64)
2	2.435 (-0.63)	2.362 (5.29)	1.507 (0.85)	2.688 (0.69)
3	2.592 (0.62)	30.993 (29.36)	16.745 (27.23)	5.878 (8.14)
1970 values:	43852	11448	18093	35807

t	S_{1t}	S_{2t}
1	1.840 (6.29)	2.066 (7.52)
2	3.123 (5.43)	3.624 (5.78)
3	5.206 (5.24)	6.523 (6.05)
1970 values:	482175	87880

TABLE 7.5

Team solution and the same Stackelberg solution for $\alpha = 0.3$

In the following tables, the optimum values of the different variables are given (we take their 1970 values to be 1.0000) and next to those quantities we give in parenthesis the corresponding yearly growth percentages (averaged over a period of 10 years).

t	x_{1t1}	x_{1t2}	x_{1t3}
1	1.504 (4.16)	1.764 (5.84)	1.701 (5.46)
2	2.575 (5.53)	2.913 (5.15)	2.814 (5.16)
3	4.490 (5.72)	3.446 (1.69)	4.027 (3.65)
1970 values:	654371	996664	2068711

t	x_{2t1}	x_{2t2}	x_{2t3}
1	1.730 (5.64)	1.417 (3.55)	2.126 (7.83)
2	2.613 (4.21)	4.720 (12.79)	4.885 (8.67)
3	2.928 (1.15)	15.170 (12.38)	10.424 (7.88)
1970 values:	253819	97018	249547

t	i^{new}_{1t1}	i^{new}_{1t2}	i^{new}_{1t3}
1	1.555 (4.51)	1.673 (5.28)	1.789 (5.99)
2	4.010 (9.94)	2.626 (4.61)	2.682 (4.13)
3	5.656 (3.50)	1.290 (-6.86)	1.989 (-2.94)
1970 values:	5503	117230	154944

t	i^{new}_{2t1}	l^{new}_{2t2}	i^{new}_{2t3}
1	2.834 (10.98)	1.352 (3.06)	2.454 (9.39)
2	1.435 (-6.58)	7.176 (18.16)	5.612 (8.62)
3	0.000 (———)	13.237 (6.31)	9.894 (5.84)
1970 values:	2651	22239	20163

t	m^{w2}_{1t1}	m^{w2}_{1t2}	m^{w1}_{2t1}	m^{w1}_{2t2}
1	2.594 (10.00)	1.411 (3.50)	1.384 (3.31)	2.509 (9.64)
2	2.719 (0.47)	1.864 (2.83)	1.672 (1.91)	3.919 (4.56)
3	2.299 (-1.66)	39.135 (35.58)	17.449 (26.43)	7.767 (7.08)
1970 values:	43852	11448	18093	35807

t	S_{1t}	S_{2t}
1	1.840 (6.29)	2.066 (7.52)
2	3.056 (5.21)	4.316 (7.65)
3	3.492 (1.34)	0.632 (7.18)
1970 values:	482175	87880

TABLE 7.6

The figures given above are only preliminary in nature and are used merely to draw some general conclusions. We first conclude that so long as DMEs and LMEs are only interested in maximizing their own consumptions, Team solution of the leader for the 3-sector model always turns out to be a Stackelberg solution as well. Furthermore, comparison of the figures in Tables 7.5 and 7.6 leads to some general observations (which are also confirmed by other simulation runs with different α's):

(1) In all the simulation runs, figures for the year 1980 are identical. Both regions have maximum savings and no overproduction. The DMEs have maximum trade deficit possible and this is effected by maximum trade deficit in the traditional sector and a slight trade surplus in the industrial sector. This is not surprising since the value-added for the LMEs are higher for the traditional and service sectors than for the industrial sector while those sectors have lower marginal capital/output coefficients; for the DMEs, on the other hand, the value-added in the industrial sector is higher than in the traditional sector, while the capital/ output coefficient for the industrial sector is lower than in the traditional sector.

(2) Difference in the figures for different α's start to appear in the year 1990. Savings are still maximum and no overproduction takes place. Trade deficit for DMEs is not maximum possible for $\alpha = 0.2$, but is again maximum for $\alpha = 0.3$. This change comes about because of increase in trade surplus for the DMEs in the industrial sector while the trade deficit in the traditional sector remains more or less the same. The effect of "Lima traget" set for the year 2000 is already noticeable in the year 1990. The LMEs have strong tendency to increase their industrial production while the DMEs pay some more attention to their traditional sector (see the growth percentages in the tables for the year 1990).

(3) In the year 2000, the effect of "Lima target" is clear. There is still no overproduction while savings for L.M.E.'s is maximum during

all the simulation runs. This is, however, not the case anymore for the DMEs when $\alpha = 0.3$. Thus for $\alpha = 0.3$, DMEs do not use their maximum growth possibilities. In all the simulations runs, the LMEs in the year 2000 have maximum trade deficit possible, effected through maximum allowed trade deficit in the traditional sector and a slight trade surplus in the industrial sector, an exact replica of the trade pattern of the DMEs in the year 1980. This is clearly the effect of imposing the "Lima target."

5. CONCLUSION

We have studied a simple world industrialization model where the world is divided into 3 regions. The centrally planned economies enter the model exogeneously, while the DMEs and LMEs play a Stackelberg game with the DMEs having the role of a leader. The DMEs and the LMEs have different linear objective functions to be minimized subject to various linear model constraints. This leads to a Stackelberg LP-problem and an algorithm for solving such problems has been discussed in the preceding chapter.

Now it is possible to formulate the problem somewhat differently. If we assume that gross output exactly matches the demand, the inequality in equation (7.1) will actually be an equality and equations (7.14)-(7.15) may be considered to give a state space description of the system. The remaining constraints may be ignored temporarily and we may consider suitable quadratic "tracking" criteria to be minimized by the DMEs and the LMEs. In formulating the respective criteria for the two regions, the constraints (7.16)-(7.19) have to be taken into account to determine the nominal trajectories and controls used in the criteria. Although the optimal solution may violate the constraints, this procedure is commonly used in economic stabilization studies (Pindyck (1973)). In fact, a similar formulation for Nash games has been suggested and solved by Pindyck (1977) in connection with a game played between the Federal Government and the Federal Reserve in the United

States while determining short term stabilization policies. Similar models for Stackelberg games have not been studied so far, but our present problem may easily be formulated and studied in that framework. The obvious advantage of this procedure is that one can give explicit recursive formulas for optimal controls (instrumental variables) for both the regions which may be computed in a rather straightforward manner. The disadvantages are numerous. The constraints may be readily violated, the state may take negative values (an impractical result) and it is very difficult to incorporate the Lima target meaningfully in the criteria functions. On the other hand, if we explicitly introduce the constraints, we are then faced with a constrained LQ-problem which does not admit simple explicit solutions.

CHAPTER 8

ECONOMETRIC MODELS : ADAPTIVE GAMES

1. INTRODUCTION

In this chapter we consider optimization of econometric models with multiple decision makers and concern ourselves with the inverse problem of estimating parameters in the model and also in the criteria under the optimal behavior of the decision makers. Since econometric models are considered, we are forced to include stochastic aspects which also play a crucial role in the parameter estimation problem. In this sense, this chapter goes beyond the general framework of our discussions so far. In accordance with the terminology in control theory, we may call the problem considered in this chapter "adaptive game". Our approach, related specifically to econometric models, is due to Chow (1981). Here one needs to use solution of linear-quadratic stochastic controls which is well-known to systems scientists. We then briefly discuss the idea of self-tuning in analogy with the corresponding treatment of adaptive control problems.

2. ESTIMATION IN LINEAR OPTIMAL CONTROL MODEL

It is well-known (see Chow (1975), for example) that linear econometric models may conveniently be represented by a linear system

$$x_k = A_k x_{k-1} + B_k u_k + c_k + n_k \tag{8.1}$$

where x_k is a state vector taking values in $\mathbb{R}^p$, u_k is a control vector taking values in $\mathbb{R}^q$, A_k, B_k, c_k are matrices and vectors of appropriate order and $\{n_k\}$ is a sequence of random vectors which are independent and identically distributed. In econometric models, one rarely starts with the state space description given by eq^n. (8.1). Most often, the relationship is given where y_k, y_{k-1}, ... , are related linearly with u_k, u_{k-1}, ... and one has to introduce state vector, often of large dimension, to express the model in the form of eq^n. (8.1). As is well-known, this is done by introducing state variables to eliminate y_{k-2}, $y_{k-3},\ldots,u_{k-1}$, u_{k-2} etc. The decision maker wants to minimize a loss function which is quadratic:

$$J = E\{ \sum_{k=1}^{N} (x_k - x_k^*)' \; Q_k (x_k - x_k^*) \} \tag{8.2}$$

where $\{x_k^*\}$ stands for reference (or desired) trajectory and Q_k is a symmetric matrix for k = 1,...,N. We did not include the control terms u_k's explicitly in the criterion. They, however, appear implicit augmenting the state space further with the incorporation of u_k as a subvector x_k. This merely simplies the apparent form of the criterion J. The solution of the optimal control problem of minimizing J in (8.2) subject to the constraint (8.1) is well-known and is given by

$$u_k = G_k \, x_{k-1} + g_k \tag{8.3}$$

where

$$G_k = -(B_k' H_k B_k)^{-1} B_k' H_k A_k \tag{8.4}$$

$$H_k = Q_k + (A_{k+1} + B_{k+1} G_{k+1})' H_{k+1} (A_{k+1} + B_{k+1} G_{k+1}) \tag{8.5}$$

$$g_k = -(B_k' H_k B_k)^{-1} B_k' (H_k c_k - h_k) \tag{8.6}$$

$$h_k = Q_k x_k^* + (A_{k+1} + B_{k+1} G_{k+1})' (h_{k+1} - H_{k+1} c_{k+1}) \tag{8.7}$$

for k = N, N-1,...,1, where

$$H_N = Q_N \quad \text{and}$$
$$h_N = Q_N x_N^* \tag{8.8}$$

We assume from now on that A, B and c are constants (independent of the time index k). We study in this section the so-called "inverse problem"; that is, estimate the parameters of the equations (8.1) and (8.2) based on the input-output data (u_k, x_k), k = 1,...,n under the assumption of optimal behavior of the controller (decision maker).

How does this problem arise in economics? Suppose that (8.1) represents a model of the national economy, u_k's are the control instruments available to the Government who wants to steer the economy along a path so as to minimize J given by (8.2). On observing the variables of the economic model and Government's decisions, econometrician may be interested in identifying the model parameters and also knowing the parameters of the Government's (quadratic) preference function. Similarly eq^n. (8.1) may represent the economic model of a private firm and (8.2) represent the criterion the firm wants to minimize. The control (or instrument) variables may be investment function, demand function for inputs etc. Econometrician may be faced with a similar problem in this case as well. In both the cases mentioned above, one is faced with the inverse problem as stated before. We assume further in this chapter that $Q_k = \beta^k Q$ and $x_k^* = F^k x^*$, where β is a discount factor and F is a diagonal matrix with diagonal elements corresponding to the time-invariant components of x_k^* being unity. Thus, our problem is to estimate the parameters A, B, c, Q and x^* based on the input-output data (u_k, x_k), $k = 1,\ldots,n$ when u_k is prescribed by eq^n. (8.3). We use the method of maximum likelihood for estimating the parameters.

We assume that we start our estimation after the system has reached the steady state. In equations (8.4)-(8.7), all the coefficients are independent of the time index k. If F = I; that is, $x_k^* = x^*$ for all k, the steady state values of G and g are found by solving the following set of algebraic equations:

$$(B'HB)G + B'HA = 0 \tag{8.9}$$

$$H - Q - \beta(A+BG)'H(A+BG) = 0 \tag{8.10}$$

$$(B'HB)g + B'(Hc-h) = 0 \tag{8.11}$$

$$[I - \beta(A+BG)']h - Qx^* + \beta(A+BG)'Hc = 0 \tag{8.12}$$

Coming back to the estimation problem, we have to evaluate the likelihood function for the system. Due to errors in implementing the optimal policy, we may assume that eq^n. (8.3) is disturbed by a noise term, the so called "residual" in the econometric literature. Thus, we have two sets of stochastic equations

$$x_k = Ax_{k-1} + Bu_k + c_k + n_k \tag{8.1}$$

and

$$u_k = Gx_{k-1} + g + v_k \tag{8.3'}$$

where we assume that $\{n_k\}$ and $\{v_k\}$ are mutually independent sequence of independent normally distributed random vectors of zero means and variances Σ and V, respectively. The joint density function of $x_1,\ldots,x_n$ and $u_1,\ldots,u_n$ is defined to be the likelihood function for the problem. Under the assumption that the residuals are normally distributed, it is reasonably straightforward to calculate the likelihood function. The logarithm of the likelihood function is given by

$$\begin{aligned}\tilde{L} = \text{constant} &- \tfrac{n}{2}\log|\Sigma| - \tfrac{n}{2}\log|V| \\ &- \tfrac{1}{2}\,\text{Tr.}[\Sigma^{-1}(X'-AX'_{-1}-BU'-cz')(X-X_{-1}A'-UB'-zc')] \\ &- \tfrac{1}{2}\,\text{Tr.}[V^{-1}(U'-GX'_{-1}-gz')(U-X_{-1}G'-zg')]\end{aligned} \tag{8.13'}$$

where $|\cdot|$ stands for the determinant of a matrix, X is a $n \times p$ matrix of observations $(x_1,\ldots,x_n)$ on the endogenous variables, X_{-1} is a $n \times p$ matrix of observations $(x_0,\ldots,x_{n-1})$ on the lagged endogenous variables, U is a $n \times q$ matrix of observations $(u_1,\ldots,u_n)'$ on the control variables, z is a vector consisting of n ones and we assume that Σ and V are nonsingular. The unknown parameters are Σ, V, A, B, c, G, H, g, h, Q and x^*. To obtain the maximum likelihood estimates of these unknown parameters, we have to maximize $\tilde{L}$ subject to the constraints (8.9)-(8.12) and an additional one: Tr. (QQ) = r which serves to normalize the matrix Q. For solving this maximization problem, we introduce the Lagrangian

$$\begin{aligned}L = \tilde{L} &- \text{Tr.}[\Omega((B'HB)G+B'HA)] - \tfrac{1}{2}\,\text{Tr.}[\Phi(H-Q-\beta(A+BG)'H(A+BG))] \\ &- \omega'[(B'HB)g+B'(Hc-h)] - \phi'[(I-(A+BG)')h-Qx^*+\beta(A+BG)'Hc] \\ &- \tfrac{1}{2}\theta[\text{Tr.}(QQ)-r]\end{aligned} \tag{8.14}$$

where Ω is a $p \times q$ matrix and Φ is a $p \times p$ symmetric matrix of Lagrange multipliers; ω a $q \times 1$ and ϕ a $p \times 1$ vector of Lagrange multipliers and

θ is a scalar Lagrange multiplier. We have to maximize L with respect to the unknown parameters which is an unconstrained problem. We use the formulas

$$\frac{\partial \log|A|}{\partial A} = A^{-1'} \quad \text{and} \quad \frac{\partial Tr.(AB)}{\partial A} = B' \tag{8.15}$$

Differentiating L with respect to the unknown parameters and setting them to zero will yield the first order conditions for optimality:

$$\frac{\partial L}{\partial \Sigma^{-1}} = n\Sigma - (X'-AX'_{-1}-BU'-cz')(X-X_{-1}A'-UB'-zc') = 0 \tag{8.16}$$

$$\frac{\partial L}{\partial V^{-1}} = nV - (U'-GX'_{-1}-gz')(U-X_{-1}G'-zg') = 0 \tag{8.17}$$

$$\frac{\partial L}{\partial A'} = X'_{-1}(X-X_{-1}A'-UB'-zc')\Sigma^{-1} - \Omega B'H$$
$$+\beta\,\Phi(A+BG)'H - \beta\phi(Hc-h)' = 0 \tag{8.18}$$

$$\frac{\partial L}{\partial B'} = U'(X-X_{-1}A'-UB'-zc')\Sigma^{-1} - \Omega'(A+BG)'H - G\Omega B'H$$
$$+\beta G\Phi(A+BG)'H - g\omega'B'H - \omega g'B'H - \omega(Hc-h)'$$
$$- \beta G\phi(Hc-h)' = 0 \tag{8.19}$$

$$\frac{\partial L}{\partial c'} = z'(X-X_{-1}A'-UB'-zc')\Sigma^{-1} - \omega'B'H$$
$$+ \beta\phi'(A+BG)'H = 0 \tag{8.20}$$

$$\frac{\partial L}{\partial G'} = X'_{-1}(U-X_{-1}G'-zg')V^{-1} - \Omega B'HB + \beta\Phi(A+BG)'HB$$
$$-\beta\phi(Hc-h)'B = 0 \tag{8.21}$$

$$\frac{\partial L}{\partial H} = -(A+BG)\Omega B' - B\Omega'(A+BG)' - \Phi + \beta(A+BG)\Phi(A+BG)'$$
$$-Bg\omega'B' - B\omega g'B' - c\omega'B' - B\omega c' - \beta c\phi'(A+BG)'$$
$$-\beta(A+BG)\phi c' = 0 \tag{8.22}$$

$$\frac{\partial L}{\partial Q} \overset{*}{=} \Phi + x^{*}\phi' + \phi x^{*'} - \theta Q \overset{*}{=} 0 \tag{8.23}$$

$$\frac{\partial L}{\partial g'} = z'(U-X_{-1}G'-zg')V^{-1} - \omega'B'HB = 0 \tag{8.24}$$

$$\frac{\partial L}{\partial h} = B\omega-[I-\beta(A+BG)]\phi = 0 \tag{8.25}$$

$$\frac{\partial L}{\partial x^{*}} \overset{*}{=} Q\phi \overset{*}{=} 0 \tag{8.26}$$

These equations follow directly except for the interpretation of the notation '$\overset{*}{=}$'. Many elements of the symmetric $p \times p$ matrix Q and of the $p \times 1$ vector x^* are usually identically zero and these are known from the actual model itself. The notation '$\overset{*}{=}$' in (8.23) and (8.26) indicates that only the derivatives of L with respect to the unknown components of Q and x^* are set equal to zero. Solution of the set of equations (8.17)-(8.26) will yield the maximum likelihood estimate of the unknown parameters of the model. This is a difficult problem and we proceed stepwise to solve this set of equations.

If there are r target variables, only a $r \times r$ submatrix of Q and r elements of x^* are nonzero. Consider equations (8.25) and (8.26). They consist of $p + r$ linear equations in $p + q$ unknowns, ϕ and ω. It is reasonable to assume that $r \geq q$, since otherwise we could have more control variables (instruments) than target variables leading to nonuniqueness of solution of our optimization problem. In this case, ϕ and ω will be zero provided the matrix

$$\begin{bmatrix} B & -[I-\beta(A+BG)] \\ 0 & Q^* \end{bmatrix}$$

has rank $p + q$, where Q^* consists of the r rows of Q corresponding to the nonzero elements of x^*. This simplifies many other equations considerably.

We first estimate (A,B,c) and (G,g) from equations (8.1) and (8.3) using the method of least squares. We use these estimates to determine Σ and V from eq^{n}s. (8.16) and (8.17). We assume, for the remaining equations, that Σ and V are known. After estimating A, B, c, G, g, from the remaining equations, we can, of course, again determine Σ and V from (8.16)-(8.17) and iterate. If we now start with any initial estimate of Q (say, a diagonal matrix with r nonzero elements whose values have been suitably guessed), we may estimate G and H by solving (8.9) and (8.10). Using eq^n, (8.9) and the fact that $\phi = 0$, we see that the last two terms of (8.21) vanish. The remaining part may be used to estimate Ω. We now use eq^n. (9.22) to determine Φ by iteration from

$$\Phi = \beta(A+BG)\Phi(A+BG)' - (A+BG)\,\Omega B' - B\Omega'(A+BG)'.$$

Eq^n. (8.23) gives us $\theta Q \overset{*}{=} \Phi$. Suppose Φ^* is composed of nonzero elements of Φ and zero elements corresponding to the zero elements of Q. Then $\Phi^* \Phi^* = \theta^2 QQ$ yields $\theta^2 = \mathrm{Tr.}(\Phi^*\Phi^*)/r$. Using this value for θ, we may estimate Q by Φ/θ. Equations (8.18) and (8.19) may now be solved to estimate A and B. Equations (8.20) and (8.24) give, respectively,

$$c = \frac{1}{n}(X'-AX'_{-1}-BU')z$$

$$g = \frac{1}{n}(U'-GX'_{-1})z$$

We thus obtain updated values for A, B, c, G, g and Q and we continue iteration until convergence.

Finally, the two remaining equations (8.11) and (8.12) are to be used to estimate h and x^*. Denote A + BG by R. We then get from (8.12)

$$k = (I-\beta R')^{-1}\,[Qx^*-\beta R'Hc]$$

and substituting this in (8.11) yields

$$B'(I-\beta R')^{-1}Qx^* = B'HBg + B'[I+\beta(I-\beta R')^{-1}R']Hc$$

Since B' is a $q \times p$ matrix, the equation above is a system of q linear equations in r unknowns in x^*. We assumed that $r \geq q$. If $r > q$, the solution for x^* is not unique. This is reasonable since the presence of the tracking term x^* in the criterion is reflected in the term g appearing in the expression for optimal control law (8.3). If x^* has more elements than g, it cannot clearly be estimated uniquely. Since we basically follow the maximum likelihood method, we expect to obtain consistent estimates of the unknown parameters by this procedure.

3. ESTIMATION IN LINEAR DIFFERENCE GAME MODEL

Inverse problem occurs more naturally when the econometrician is confronted with a difference game model. In this case there are multiple decision makers and we restrict ourselves only to the situation of two economic decision makers. The econometric model may now be represented by

$$x_k = Ax_{k-1} + B_1u_{1k} + B_2u_{2k} + c_k + n_k \tag{8.27}$$

where x_k is a state vector in $\mathbb{R}^p$, u_{1k} and u_{2k} are control vectors of the two economic decision makers taking values in $\mathbb{R}^{q_1}$, $\mathbb{R}^{q_2}$, A, B_1, B_2 are constant matrices of appropriate dimensions and $\{u_k\}$ are a sequence of random vectors which are independent and identically distributed. DM i chooses its control variable u_{ik} so as to minimize a quadratic cost function

$$E\{\sum_{k=1}^{N}(x_k-x^*_{ik})'Q_{ik}(x_k-x^*_{ik})\},\ i = 1,2 \tag{8.28}$$

where $\{x^*_{ik}\}$ stands for the reference trajectory and Q_{ik} is a symmetric matrix for $k = 1,\dots,N$; $i = 1,2$. The optimum decision of DM i on the assumption that the control of the other DM is fixed is given by

$$u_{ik} = G_{ik}\,x_{k-1} + g_{ik},\ i = 1,2 \tag{8.29}$$

As in the previous section, the econometrician's problem is to estimate the parameters of (8.27) and (8.28) based on the input-output data (x_k, u_{1k}, u_{2k}), $k = 1,\dots,n$. When one of the DMs is the government, we may be interested in evaluating the effects of government policy changes and the choice of an optimum decision for the government.

One can think of several dynamic games in economics where this type of inverse problem is relevant. u_{1k} may represent control variables of some group of decision makers of the private sector and u_{2k} may represent controls available to the government. Alternatively, u_{1k} and u_{2k} may represent control variables of the fiscal (ministry of finance) and monetary (central bank) authorities of the central government (see Pindyck (1977)). Economic game between two countries (or, groups of countries) where each player wants to estimate the parameters chosen by the other player for his model and optimizing criterion also leads to realistic and interesting inverse problem in dynamic games.

It is obvious that the dynamic game model considered here means that the decision of one player will affect the equation (environment) used by the other player to optimize his criterion and vice versa. Thus, in the example of dynamic game between a group of DMs in the private sector and the government, not only does the government decision rule $u_{2k} = G_{2k}x_{k-1} + g_{2k}$ affect the optimal policy of the private sector, the latter's optimal decision rule $u_{1k} = G_{1k}x_{k-1} + g_{1k}$ will also affect the decision rule of the government. We have two different situations depending on whether the two players play noncooperatively (Nash game) or one of the players, say, the government, has the role of a dominant player (Stackelberg game). We discuss both the cases below, although Stackelberg game is treated more elaborately.

We first consider the estimation problem when the DM2 (the government in our first example in the preceding paragraph) is the dominant player of the dynamic game.

Thus we assume that the government adheres to a decision rule $u_{2k} = G_{2k}\,x_{k-1} + g_{2k}$. Then the private sector has to optimize eq^n. (9.28) for $i = 1$ subject to (8.27) and the constraint imposed by the government. The government then takes this reaction of DM1 into account to optimize (8.28) for $i = 2$ and thereby determines G_{2k} and g_{2k}. In this process, we also obtain decision rule $u_{1k} = G_{1k}\,x_{k-1} + g_{1k}$ of DM1. We now assume that the estimation process starts when the system has reached the steady state so that G_i and g_i, $i = 1,2$, are constants (independent of k). We consider first the simple case where we do not estimate Q_{2k} and x^*_{2k}, assuming them to be known. Adding residuals v_{ik} to (8.29) as in section 2, we could furthermore estimate G_2 and g_2 by treating them as coefficients of a least-squares regression of u_{2k} on x_{k-1}. Once G_2 and g_2 are given, optimizing (8.28) for $i = 1$ will yield G_1 and g_1 as functions of the parameters of (8.27) and Q_{1k}, x^*_{1k} of (8.28). We assume now, as in the previous section, that n_k, v_{1k} and v_{2k} are jointly normally distributed so that one can easily write down the likelihood function for the problem which depends on the parameters of (8.27) and on G_i, g_i, $i = 1,2$, appearing in (8.29). In view of our preceding discussion, the likelihood function for the simple case considered now will depend on the system parameters and also on Q_{1k} and x^*_{1k}. Assume now that $Q_{1k} = \beta_1^k Q_{10}$ and $x^*_{1k} = F_1^k x^*_1$, β_1 being a discount factor for the private sector and F_1 being a diagonal matrix with some elements known to be one if the targets in x^*_{1k} are constants. The likelihood function is, therefore, to be maximized with respect to the system parameters and Q_{10}, β_1, x^*_1 and F_1. This maximization can be carried out as in the previous section.

The case of unknown Q_{2k} and x^*_{2k} which are to be estimated is considerably more difficult. We treat this case under the additional assumption that c_k, Q_{1k}, x^*_{1k}, Q_{2k} and x^*_{2k} are all independent of k (time-invariant). In this case, we explicitly determine (G_1,g_1) and (G_2,g_2) given the system parameters and Q_1, x^*_1, Q_2, x^*_2. Then, given the system parameters and Q_1, x^*_1, Q_2, x^*_2, we are able to evaluate the likelihood function. One can then use a gradient method to maximize the likelihood function with respect to these parametars and therefore determine the maximum likelihood estimates of all the unknown parameters appearing in the system equation and the optimizing criteria. Explicit determination of (G_1,g_1) and (G_2,g_2) proceeds as follows. Suppose that DM2 selects a feedback strategy $u_{2k} = G_2 x_{k-1} + g_2$. State equation confronting DM1 then becomes

$$x_k = (A+B_2G_2)x_{k-1} + B_1u_{1k} + c + B_2g_2 + n_k \qquad (8.27')$$

and DM1 wants to minimize the criterion

$$E \{ \sum_{k=1}^{N} (x_k-x^*_1)'Q_1(x_k-x^*_1)\} \qquad (8.28')$$

We have already seen in section 2 (equations (8.9)-(8.12)) that the steady state strategy for DM1 is $u_{1k} = G_1x_{k-1} + g_1$, where

$$(B_1'HB_1)G_1 + B_1'H_1(A+B_2G_2) = 0 \qquad (8.30)$$

$$H_1 - Q_1 - (A+B_2G_2+B_1G_1)'H_1(A+B_2G_2+B_1G_1) = 0 \tag{8.31}$$

$$(B_1'H_1B_1)g_1 + B_1'[H_1(c+B_2g_2)-h_1] = 0 \tag{8.32}$$

$$[I-(A+B_2G_2+B_1G_1)']h_1 - Q_1\overset{*}{x}_1 + (A+B_2G_2+B_1G_1)'H_1(c+B_2g_2) = 0 \tag{8.33}$$

Given G_2, eq^{n}s. (8.30)-(8.31) can be solved to determine G_1 and H_1. Furthermore, given g_2, eq^{n}s. (8.32)-(8.33) can be solved to obtain g_1 and h_1. Suppose now that in the steady state, the state vector has mean $\bar{x}$ and covariance $\Gamma = E(x_k-\bar{x})(x_k-\bar{x})'$. Writing $u_{1k} = G_1x_{k-1} + g_1$ in (8.27'), we readily see that $\bar{x}$ and Γ satisfy

$$(I-A-B_1G_1-B_2G_2)\bar{x} - c - B_1g_1 - B_2g_2 = 0 \tag{8.34}$$

$$\Gamma - (A+B_1G_1+B_2G_2)T(A+B_1G_1+B_2G_2)' - En_kn_k' = 0 \tag{8.35}$$

We now consider the optimization problem of the DM2. It is the problem of minimizing, in the stationary state,

$$\frac{1}{2} E\{(x_k-\overset{*}{x}_2)'Q_2(x_k-\overset{*}{x}_2)\} = \frac{1}{2} \text{Tr. } (Q_2\Gamma) + \frac{1}{2} (\bar{x}-\overset{*}{x}_2)'Q_2(\bar{x}-\overset{*}{x}_2)$$

with respect to G_2 and g_2 subject to the constraints (8.30)-(8.35). This problem can be solved by introducing the Lagrangian

$$\begin{aligned} L = {} & \frac{1}{2} \text{Tr.}(Q_2\Gamma) + \frac{1}{2}(\bar{x}-\overset{*}{x}_2)'Q_2(\bar{x}-\overset{*}{x}_2) - \omega'[B_1'H_1B_1g_1+B_1'(H_1(c+B_2g_2)-h_1)] \\ & - \phi'[(I-(A+B_2G_2+B_1G_1)')h_1 - Q_1\overset{*}{x}_1 + (A+B_2G_2+B_1G_1)'H_1(c+B_2g_2)] \\ & - \lambda'[(I-A-B_1G_1-B_2G_2)\bar{x} - (c+B_1g_1+B_2g_2)] \\ & - \text{Tr.}[\Omega(B_1'HB_1G_1+B_1'H_1(A+B_2G_2))] \\ & - \frac{1}{2} \text{Tr. } [\Phi(H_1-Q_1-(A+B_2G_2+B_1G_1)'H_1(A+B_2G_2+B_1G_1))] \\ & - \frac{1}{2} \text{Tr.}[\Psi(\Gamma-(A+B_1G_1+B_2G_2)\Gamma(A+B_1G_1+B_2G_2)' - En_kn_k')] \end{aligned} \tag{8.36}$$

where ω, ϕ, λ, Ω, $\Phi = \Phi'$ and $\Psi = \Psi'$ are vectors and matrices of Lagrange multipliers. We use the differentiation formula ∂ Tr. $(AB)/\partial A = B'$ and write R to denote $A + B_1G_1 + B_2G_2$. Differentiating L with respect to the unknown parameters and setting them to zero will yield the first order conditions for optimality:

$$\frac{\partial L}{\partial g_1} = -B_1'H_1B_1\omega + B_1'\lambda = 0 \tag{8.37}$$

$$\frac{\partial L}{\partial h_1} = B_1\omega - (I-R)\phi = 0 \tag{8.38}$$

$$\frac{\partial L}{\partial g_2} = - B_2'H_1B_1\omega - B_2'H_1R\phi + B_2'\lambda = 0 \tag{8.39}$$

$$\frac{\partial L}{\partial \bar{x}} = Q_2(\bar{x}-x_2^*) - (I-R')\lambda = 0 \tag{8.40}$$

$$\frac{\partial L}{\partial G_1} = B_1'[-H_1B_1\Omega'+H_1R\Phi + \Psi R\Gamma + h_1\phi' - H_1(c+B_2g_2)\phi' + \lambda\bar{x}'] = 0 \tag{8.41}$$

$$\frac{\partial L}{\partial G_2} = B_2' [-H_1B_1\Omega'+H_1R\Phi+\Psi R\Gamma+h_2\phi' - H_1(b+B_2g_2)\phi'+\lambda\bar{x}'] = 0 \tag{8.42}$$

$$\begin{aligned}\frac{\partial L}{\partial H_1} = &- \Phi + R\Phi R' - B_1g_1\omega'B_1' - B_1\omega g_1'B_1' \\ &- (c+B_2g_2)\omega'B_1' - B_1\omega(c+B_2g_2)'-(c+B_2g_2)\phi'R-R'\phi(c+B_2g_2)'=0\end{aligned} \tag{8.43}$$

$$\frac{\partial L}{\partial \Gamma} = Q_2 - \Psi + R'\Psi R = 0 \tag{8.44}$$

To solve these equations (8.37)-(8.44) are extremely difficult. We do this in two stages. First consider a simpler problem of minimizing Tr.$(Q_2\Gamma)$ with $c = 0$, $x_1^* = 0$ and $x_2^* = 0$. The optimal strategies are then $u_{1k} = G_1x_{k-1}$ and $u_{2k} = G_2x_{k-1}$, the constraints (8.32), (8.33) and (8.34) have no relevance and one needs to solve equations (8.41)-(8.44) with $\omega = 0$, $\phi = 0$, $\lambda = 0$, $\bar{x} = 0$ and $g_2 = 0$. Eq^n. (8.43) then becomes $\Phi = R\Phi R'$ which has $\Phi = 0$ as a solution. Eq^n. (8.41) gives us

$$\Omega' = (B_1'H_1B_1)^{-1} B_1' \Psi R\Gamma$$

which, when substituted in (9.42), yields

$$B_2'[I-H_1B_1(B_1'H_1B_1)^{-1}B_1']\Psi(A+B_1G_1+B_2G_2)\Gamma = 0 \tag{8.45}$$

Now, starting with an initial guess for G_2, we can solve (8.30) and (8.31) to determine G_1 and H_1. Given G_1, we solve eq^n. (8.44) for Ψ. Then eq^n. (8.45), expressed in the form (on postmultiplying by Γ^{-1})

$$\begin{aligned}G_2 = &\{B_2'[I-H_1B_1(B_1'H_1B_1)^{-1}B_1']\Psi B_2\}^{-1}x \\ &B_2'[I-H_1B_1(B_1'H_1B_1)^{-1}B_1']\Psi(A+B_1G_1)\end{aligned}$$

can be used to compute new G_2. This iterative process can be continued until convergence to determine G_2, G_1 and H_1 for the simpler problem. Let us now consider the general problem. With P_1 denoting $B_1(B_1'H_1B_1)^{-1}B_1'$, equations (8.37)-(8.40) give us

$$\omega = (B_1'H_1B_1)^{-1}B_1'\lambda \tag{8.37'}$$

$$\phi = (I-R)^{-1}P_1\lambda \tag{8.38'}$$

$$B_2'\{I-H_1[I+R(I+R(I-R)^{-1}]P_1\}\lambda = 0 \tag{8.39'}$$

$$\lambda = (I-R')^{-1}Q_2(\bar{x}-x_2^*) \tag{8.40'}$$

Eq^{n}s. (8.40') and (8.34) give us

$$\lambda = (I-R')^{-1}Q_2[(I-R)^{-1}(c+B_1g_1+B_2g_2)-x_2^*] \tag{8.46}$$

Combining (9.46) with (9.39') gives

$$B_2'\{I-H_1[I+R(I-R)^{-1}]P_1\}(I-R')^{-1}Q_2[(I-R)^{-1}(c+B_1g_1+B_2g_2)-x_2^*] = 0 \tag{8.47}$$

Using the approach indicated for the simpler problem, we start with the preceding approximate values for G_2, G_1 and H_1. Then (8.47) along with (8.32) and (8.33) can be solved for g_2, g_1 and h_1. This can be done by expressing g_2 as a linear function of g_1 in eq^n, (8.47) and using this, (8.32)-(8.33) become two linear equations in g_1 and h_1. Equations (8.46), (8.37') and (8.38') may then be used to find λ, ω and ϕ and finally, (8.34) may be used to determine $\bar{x}$.

We now follow the steps of the simpler problem to determine new G_2, G_1 and H_1. We use eq^n. (8.43) to determine Φ iteratively. Equations (8.41) and (8.42) imply

$$\Omega' = (B_1'H_1B_1)^{-1}B_1'[H_1R\Phi+\Psi R\Gamma+\ldots] \tag{8.41'}$$

$$B_2'[I-H_1P_1][H_1R\Phi+\Psi R\Gamma+h_2\phi'-H(b+B_2g_2)\phi'+\lambda\bar{x}'] = 0 \tag{8.42'}$$

Equations (8.35) and (8.44) may be used to compute Γ and Ψ respectively. Postmultiplying (8.42') by Γ^{-1}, we get the relation

$$B_2'[I-H_1P_1]\Psi B_2G_2 = B_2'[I-H_1P_1][H_1R\Phi\Gamma^1+\Psi(A+B_1G_1)+\ldots]$$

which can be used to determine G_2 recursively. This enables us to determine G_1 and H_1 from (8.30)-(8.31). Having obtained the new set of G_2, G_1 and H_1, we can go back to determining g_2, g_1 and h_1 and continue this iterative process until convergence.

If the decision makers play noncooperatively, or a Nash game, the estimation problem is simpler. First, assume that the decision rule of the government is known; that is, (G_2, g_2) is given. With $Q_{1k} = \beta_1^k Q_{10}$ and $x_{1k}^* = F_1^k x_1^*$, we can use the method of section 2, under the assumption of optimal behavior for the private sector, to estimate the parameters of (8.27) and Q_{10}, β_1 x_1^*, F_1 and thereby, to estimate G_1 and g_1. Next, assume that the private sector uses the decision rule specified by (G_1, g_1). With $Q_{2k} = \beta_2^k Q_{20}$ and $x_{2k}^* = F_2^k x_2^*$, we can similarly estimate the parameters of (8.27) and Q_{20}, β_2, x_2^*, F_2 and thereby, G_2 and g_2. We can now go back to our first step and iterate until convergence. If we assume that Q_{ik} and x_{ik}^*, $i = 1,2$, are independent of k (time invariant), we can maximize the likelihood function subject to two sets of constraints. One set consists of equations (8.30)-(8.33) and the other one is identical except for the subscripts 1 and 2 interchanged.

4. CONCLUDING REMARKS

Various estimation procedures discussed above depend on iterative methods and work satisfactorily only when the iterations converge. If the iterations show divergence, one has reason to suspect the assumption that each player reacts optimally to the decision of the other and the effort of estimating the parameters in our present set-up should be abandoned. It may also be used as a test regarding the rational behavior of the players. Numerical experiences of the methods outlined in section 3 are still very limited. We have also separated the optimization and estimation problems to solve the adaptive game and explicit convergence results of the parameters in this context are not available yet. Inverse problem discussed in section 3 may also be useful for a "post data" analysis to see whether the players indeed played a Nash game or that one of the players had a dominant role in a Stackelberg game. Although a question of considerable practical importance, this has not been analyzed so far.

Finally, we note that the estimation procedure discussed above is "off-line" because the input-output data (u_{1k}, u_{2k}, x_k), $k = 1,\ldots,n$, is fixed. In many practical situations, one may be interested in knowing how the estimate changes with n; that is, with the availability of increasing amount of data. Estimation procedures which explicitly indicate this dependence on the available data, and the update thereof, are called "on-line". A special problem of this type has been studied by Chan and Cruz (1983). They consider an adaptive Nash game with decentralized cost criteria and a special type of ARMA model. ARMA model is, of course, quite natural in econometrics. What makes their model restrictive is that the state may be partitioned into two components with each having exactly the same number of components as the number of control variables of each decision maker. The criteria are decentralized in the sense that they involve quadratic tracking terms for i-th player involving only the i-th control vector and the corresponding i-th"block" component of the state vector for $i = 1,2$. Certainty equivalence is imposed on the system and estimates of the parameters are substituted into the optimal decision rules. Since "on-line" results are sought for, a recursive identification scheme due to Ljung (1979) is used for estimating the parameters. One needs to solve the Nash game for the case of known parameters. The substitution in this solution of the recursive scheme mentioned above for the parameters results in the desired algorithm for solving the problem. The authors solve the problem under the so-called one-step delay observation sharing pattern. The algorithm is then applied to a very simple economic model with two inputs and two outputs. The inputs are government expenditure under control of the ministry of finance and money supply under control of the central bank. The outputs are the consumption expenditure and private investment. To apply the model, the authors assume that the government wants to keep only the consumption expenditure close to a desired trajectory while the central bank wants to keep only the money supply close to a desired target path. This is, of course, not quite realistic. Convergence analysis of this adaptive scheme is also an open question. Simulation studies show that the algorithm is capable of stabilizing the system along desired trajectories asymptotically. Because of the complete separation between optimization and estimation, extension of this method to Stackelberg games is not difficult, although it has not yet been carried out in detail.

REFERENCES

Arrow, K.J. and M. Kurz (1970), Public Investment, the Rate of Return and Optimal Fiscal Policy, John Hopkins Press, Baltimore, MD.

Bagchi, A. and T, Başar (1981), "Stackelberg Strategies in Linear-Quadratic Stochastic Differential Games", Journal of Optimization Theory and Applications, 35, no. 3,443 - 464.

Bagchi, A., G.J. Olsder, and R.C.W. Strijbos (1981), "Regional Allocation of Investment as a Hierarchical Optimization Problem", Regional Science and Urban Economics, 11, no. 2, 205 -213.

Bagchi, A., M. Moraal, and G.J. Olsder (1982), "Optimization in Hierarchical Setting of a Simple World Industrialization Model". In R.F. Drenick and F. Kozin, eds., System Modeling and Optimization, Lecture Notes in Control and Information Sciences, no 38, 621 - 629, Springer-Verlag, Berlin.

Balch, M. (1971), "Oligopolies, Advertising and Noncooperative Games". In H.W. Kuhn and G.P. Szego, eds., Differential Games and Related Topics, 301 - 311, North-Holland, Amsterdam.

Başar, T. (1977), "Informationally Nonunique Equilibrium Solutions in Differential Games", SIAM Journal on Control and Optimization, 15, no. 4, 636 -660.

Başar, T. and G.J. Olsder (1982), Dynamic Noncooperative Game Theory, Academic Press, London.

Billera, L.J. (1979), "Economic Market Games." In W.F. Lucas, ed., Game Theory and Its Applications, 37-53, American Mathematical Society, Providence, RI.

Boltyanski, V.G. (1978), Optimal Control of Discrete Systems, Hallsted Press, John Wiley, New York.

Brock, W.A. (1971), "Sensitivity of Optimal Growth Paths with respect to a Change in Target Stocks", Zeitschrift für Nationalökonomie, Supp. 1

Case, J.H. (1971), "Applications of the Theory of Differential Games to Economic Problems". In H.W. Kuhn and G.P. Szego, eds., Differential Games and Related Topics, 345-371, North-Holland, Amsterdam.

Case, J.H. (1979), Economics and The Competitive Process, New York University Press, New York.

Cesari, L. (1983), Optimization:Theory and Applications, Springer-Verlag, New York.

Chakravarti, S. (1969), Capital and Development Planning, MIT Press, Cambridge, MA.

Chan, Y.M. and J.B. Cruz, Jr (1983), "Decentralized Stochastic Adaptive Nash Games", Optimal Control Applications & Methods, 4, 163-178.

Chen, C.I. and J.B. Cruz, Jr (1972), "Stackelberg Solution for Two-Person Games with Biased Information Patterns", IEEE Transactions on Automatic Control, AC-17, no. 6, 791-797.

Chow, G.C. (1975), Analysis and Control of Dynamic Economic System, John Wiley, New York.

Chow, G.C. (1981), Econometric Analysis by Control Methods, John Wiley, New York.

Das, R.K. (1974), Optimal Investment Planning, Rotterdam University

Press, Rotterdam.

Dorfman, R. (1969), "An Economic Interpretation of Optimal Control Theory", The American Economic Review, 59, no. 2, 817 - 831.

Feichtinger, G., ed. (1982), Optimal Control Theory and Economic Analysis North-Holland, Amsterdam.

Fleming, W. H. and R.W. Rishel (1975), Deterministic and Stochastic Optimal Control, Springer-Verlag, Berlin.

Friedman, A. (1971), Differential Games, John Wiley, New York.

Friedman, J.W. (1977), Oligopoly and the Theory of Games, North-Holland, Amsterdam.

Fujita, M. (1978), Spatial Development Planning : A Dynamic Convex Programming Approach, North-Holland, Amsterdam.

Halkin, H. (1974), "Necessary Conditions for Optimal Control Problems with Infinite Horizons", Econometrica, 42, no. 2, 267 -272.

Hauptmann, H. (1982), "Stackelberg Strategies and An Embargo Game", In G. Feichtinger, ed., Optimal Control Theory and Economic Analysis, 337-351, North-Holland, Amsterdam.

Hestenes, M.R. (1965), "On Variational Theory and Optimal Control Theory", SIAM Journal on Control, 3, 23 - 48.

Ho, Y.C., P.B. Luh, and G.J. Olsder (1980), "A Control-Theoretic View on Incentives". In A. Bensoussan and J.L. Lions, eds., Analysis and Optimization of Systems, Lecture Notes in Control and Information Sciences, no. 28, 359-383, Springer-Verlag, Berlin.

Intrilligator, M.S. (1964), "Regional Allocation of Investment:Comment", Quarterly Journal Of Economics, 73.

Isaacs, R. (1975), Differential Games, 2nd edn., Kruger Publishing Company, Huntington, New York.

Isard, W. And P. Liossatos (1979), Spatial Dynamics and Optimal Space-Time Development, North-Holland, Amsterdam.

Jørgensen, S. (1982), "A Survey of Some Differential Games in Advertising", Economic Dynamics and Control, 4, no. 4, 341 - 369.

Kamien, M.I. and N.L. Schwartz (1971), "Sufficient Conditions in Optimal Control Theory", Journal of Economic Theory, 3, 207 - 214.

Kendrick, D. (1981), Stochastic Control for Economic Models, McGraw Hill, New York.

Kuhn, H.W. (1953), "Extensive Games and The Problem of Information". In H.W. Kuhn and A.W. Tucker, eds., Contributions to the Theory of Games, 193 -216, Annals of Mathematics Studies, no. 28, Princeton University Press, Princeton, NJ.

Kuilder, H. (1982), "Investment Reallocation in a Control-Theoretic Setting", Master's Thesis, Twente University of Technology, Enschede, Netherlands (in Dutch).

Ljung, L. (1979), "Convergence of an Adaptive Filter Algorithm", International Journal on Control, 27, no. 5, 673 - 693.

Mangasarian, O.L. (1966), "Sufficient Conditions for the Optimal Control of Nonlinear Systems", SIAM Journal on Control, 4, 139 - 152.

Marschak, J. and R. Radner (1972), Economic Theory of Teams, Yale University Press, New Haven, CT.

Mehlmann, A. (1982), "On Relations between Open-Loop and Closed-loop Nash Solutions in Deterministic Differential Games". In G. Feichtinger, ed., Optimal Control Theory and Economic Analysis, 399 -413, North-Holland, Amsterdam.

Moraal, M. (1980), "State Space Representation and Simulation of a Simple World Industrialization Model", TW Memorandum, no. 300, Department of Applied Mathematics, Twente University of Technology, Enschede, Netherlands.

Moraal, M. (1981), "Stackelberg Solutions in Linear Programming Problems", Methods of Operations Research, 44, 375 - 383.

Nash, J. (1951), "Noncooperative Games", Annals of Mathematics, 54 no. 2, 286-295.

Von Neumann, J. and O. Morgenstern (1947), Theory of Games and Economic Behavior, 2nd edn., Princeton University Press, Princeton, NJ.

Nygård, F. and A. Sandström (1981), Measuring Income Inequality Almquist and Wiksell International, Stockholm.

Oniki, H. (1973), "Comparative Dynamics (Sensitivity Analysis) in Optimal Control Theory", Journal of Economic Theory, 6, no. 3, 265 - 283.

Opdam, H. and A. Ten Kate (1978), "A simple World Industrialization Model", Internal Report, Centre for Development Planning, Erasmus University, Rotterdam.

Owen, G. (1968), Game Theory, Saunders, Philadelphia, PA.

Papavassilopoulos, G.P. and J.B. Cruz, Jr (1979), "Nonclassical Control Problems and Stackelberg Games", IEEE Transactions on Automatic Control, AC-24, no. 2, 155 -166.

Pindyck, R.S. (1973), Optimal Planning for Economic Stabilizations, North-Holland, Amsterdam.

Pindyck, R.S. (1977), "Optimal Planning for Economic Stabilization Policies under Decentralized Control and Conflicting Objectives", IEEE Transactions on Automatic Control, AC-22, no. 4, 517-530.

Pitchford, J.D. (1977), "Two State Variable Problems". In J.D. Pitchford and S.J. Turnovsky, eds., Applications of Control Theory to Economic Analysis, 127-154, North-Holland, Amsterdam.

Pitchford, J.D. and S.J. Turnovsky, eds. (1977), Applications of Control Theory to Economic Analysis, North-Holland, Amsterdam,

Rahman, M.A. (1963), "Regional Allocation of Investment", Quarterly Journal of Economics, 72, no. 1, 26-39.

Rahman, M.A. (1966), "Regional Allocation of Investment:Continuous Version", Quarterly Journal of Economics, 75.

Ramsey, F.P. (1928), "A Mathematical Theory of Saving", Economic Journal, 38, 543 - 559.

Reinganum, J.F. (1983), "Technology Adoption under Imperfect Information", The Bell Journal of Economics, 14, no. 1, 57 -69.

Roy, J.R. and P.F. Lesse (1983), "Planning Models for Noncooperative Situations : A Two-Player Game Approach", Regional Science and Urban Economics, 13, no. 2, 195 - 211.

Sandell, N.R. Jr (1974), "On Open-Loop and Closed-Loop Nash Strategies", IEEE Transactions on Automatic Control, AC-19, no. 4. 435 - 436.

Sen, A.K. (1976), "Poverty : An Ordinal Approach to Measurement", Econometrica, 44, no. 2, 219 -231.

Shell, K., ed. (1967), Essays on the Theory of Optimal Economic Growth, MIT Press, Cambridge, MA.

Shubik, M. (1975), Games for Society, Business and War, Elsevier, Amsterdam.

Simaan, M. and J.B. Cruz, Jr (1973), "On the Stackelberg Strategy in Nonzero-sum Games", Journal of Optimization Theory and Applications, 11, no. 5, 533 - 555.

Simaan, M. and T. Takayama (1978), "Game Theory applied to Dynamic Duopoly Problems with Production Constraints", Automatica, 14, 161 - 166.

Von Stackelberg, H. (1934), Marktform und Gleichgewicht, Springer, Vienna (English Translation : The Theory of the Market Economy, Oxford University Press, Oxford, 1952).

Starr, A.W. and Y.C. Ho (1969), "Nonzero-sum Differential Games", Journal of Optimization Theory and Applications, 3, no. 3, 184-206.

Strijbos, R.C.W. and A. Bagchi (1982), "Regional Allocation of Investment : New Problems in Hierarchical Control", Large Scale Systems, 3, 89-95.

Takayama, A. (1974), Mathematical Economics, The Dryden Press, Hindsdale, Illinois.

Togsverd, T. (1975), "Multilevel Planning in the Public Sector". In H. Thiriez and S. Zionts, eds., Multiple Criteria Decision Making, Lecture Notes in Economics and Mathematical Systems, no. 201-214, Springer-Verlag, Berlin.

Vorob'ev, N.H. (1977), Game Theory, Springer-Verlag, Berlin.

Weber, R.J. (1979), "Noncooperative Games". In W.F. Lucas, ed., Game Theory and Its Applications, 83 - 125, American Mathematical Society, Providence, RI.

Galbraith, J.K. (1973), Economics and The Public Purpose, Houghton Mifflin Company, Boston,

Lecture Notes in Control and Information Sciences

Edited by A. V. Balakrishnan and M. Thoma

Vol. 43: Stochastic Differential Systems
Proceedings of the 2nd Bad Honnef Conference of the SFB 72 of the DFG at the University of Bonn
June 28 – July 2, 1982
Edited by M. Kohlmann and N. Christopeit
XII, 377 pages. 1982.

Vol. 44: Analysis and Optimization of Systems
Proceedings of the Fifth International Conference on Analysis and Optimization of Systems
Versailles. December 14–17, 1982
Edited by A. Bensoussan and J. L. Lions
XV, 987 pages, 1982

Vol. 45: M. Arató
Linear Stochastic Systems with Constant Coefficients
A Statistical Approach
IX, 309 pages. 1982

Vol. 46: Time-Scale Modeling of Dynamic Networks with Applications to Power Systems
Edited by J. H. Chow
X, 218 pages. 1982

Vol. 47: P. A. Ioannou, P. V. Kokotovic
Adaptive Systems with Reduced Models
V, 162 pages. 1983

Vol. 48: Yaakov Yavin
Feedback Strategies for Partially Observable Stochastic Systems
VI, 233 pages, 1983

Vol. 49: Theory and Application of Random Fields
Proceedings of the IFIP-WG 7/1 Working Conference held under the joint auspices of the Indian Statistical Institute
Bangalore, India, January 1982
Edited by G. Kallianpur
VI. 290 pages. 1983

Vol. 50: M. Papageorgiou
Applications of Automatic Control Concepts to Traffic Flow Modeling and Control
IX, 186 pages. 1983

Vol. 51: Z. Nahorski, H.F. Ravn, R.V.V. Vidal
Optimization of Discrete Time Systems
The Upper Boundary Approach
V, 137 pages 1983

Vol. 52: A. L. Dontchev
Perturbations, Approximations and Sensitivity Analysis of Optimal Control Systems
IV, 158 pages. 1983

Vol. 53: Liu Chen Hui
General Decoupling Theory of Multivariable Process Control Systems
XI, 474 pages. 1983

Vol. 54: Control Theory for Distributed Parameter Systems and Applications
Edited by F. Kappel, K. Kunisch, W. Schappacher
VII, 245 pages. 1983.

Vol. 55: Ganti Prasada Rao
Piecewise Constant Orthogonal Functions and Their Application to Systems and Control
VII, 254 pages. 1983.

Vol. 56: Dines Chandra Saha, Ganti Prasada Rao
Identification of Continuous Dynamical Systems
The Poisson Moment Functional (PMF) Approach
IX, 158 pages. 1983.

Vol. 57: T. Söderström, P. G. Stoica
Instrumental Variable Methods for System Identification
VII, 243 pages. 1983.

Vol. 58: Mathematical Theory of Networks and Systems
Proceedings of the MTNS-83 International Symposium
Beer Sheva, Israel, June 20–24, 1983
Edited by P. A. Fuhrmann
X, 906 pages. 1984

Vol. 59: System Modelling and Optimization
Proceedings of the 11th IFIP Conference
Copenhagen, Denmark, July 25-29, 1983
Edited by P. Thoft-Christensen
IX, 892 pages. 1984

Vol. 60: Modelling and Performance Evaluation Methodology
Proceedings of the International Seminar
Paris, France, January 24–26, 1983
Edited by F. Bacelli and G. Fayolle
VII, 655 pages. 1984

Vol. 61: Filtering and Control of Random Processes
Proceedings of the E.N.S.T.-C.N.E.T. Colloquium
Paris, France, February 23–24, 1983
Edited by H. Korezlioglu, G. Mazziotto, and J. Szpirglas
V, 325 pages. 1984

Lecture Notes in Control and Information Sciences

Edited by A. V. Balakrishnan and M. Thoma

Vol. 62: Analysis and Optimization
of Systems
Proceedings of the Sixth International
Conference on Analysis and Optimization
of Systems
Nice, June 19–22, 1984
Edited by A. Bensoussan, J. L. Lions
XIX, 591 pages. 1984.

Vol. 63: Analysis and Optimization
of Systems
Proceedings of the Sixth International
Conference on Analysis and Optimization
of Systems
Nice, June 19–22, 1984
Edited by A. Bensoussan, J. L. Lions
XIX, 700 pages. 1984.

Vol. 64: Arunabha Bagchi
Stackelberg Differential Games
in Economic Models
VIII, 203 pages, 1984